Reading with Michel Serres

SUNY Series, The Margins of Literature
Mihai I. Spariosu, editor

Reading with Michel Serres

An Encounter with Time

Maria L. Assad

STATE UNIVERSITY OF NEW YORK PRESS

Published by
State University of New York Press, Albany

Printed in the United States of America

For information, address
State University of New York Press, State
University Plaza, Albany, N.Y., 12246

Production by Dale Cotton
Marketing by Nancy Farrell

Library of Congress Cataloging-in-Publication Data

Assad, Maria L., 1938–
Reading with Michel Serres : an encounter with time / Maria L. Assad.
p. cm.
Includes bibliographical references and index.
ISBN 0-7914-4229-2 (hc : alk. paper). — ISBN 0-7914-4230-6 (pb : alk. paper)
1. Serres, Michel Contributions in philosophy of time. 2. Time.
I. Title.
B2430.S46A88 1999
115′.092—dc21 98-27740
CIP

10 9 8 7 6 5 4 3 2 1

Contents

To the memory of my brothers
Hans and Herbert

Acknowledgments

The inspiration for this study came originally from a series of lectures by Michel Serres which I attended. They reawakened in me a latent interest in scientific questions which mingled with what I was listening to in such a way that something new emerged, something with self-organizing tendencies that grew into a thought construct of its own. For this passage of discovery I thank Michel Serres.

I also wish to thank my friends and colleagues of the Society for Literature and Science (SLS) for their support, encouragement, and helpful suggestions while I was working on this project. My special thanks go to Thomas Weissert who read several sections of the manuscript and clarified many of the questions I had on complex dynamics, and did so with unwavering patience and understanding. The comments and critical evaluations by the anonymous readers at State University of New York Press were of great assistance to me while I was engaged in final revisions. I much appreciate their careful readings. I thank especially my editor at State University of New York Press, James Peltz, without whose consistent and dedicated help this project would not have materialized. Any shortcomings this work may have are certainly not due to the efforts of all these individuals, but are mine and mine alone.

Finally, to my husband go my thanks for his patience while my study was in progress. These few words cannot adequately express my grateful appreciation for his unfailing support and understanding whenever he found me with my nose in my books.

Acknowledgments

Excerpts from Michel Serres's *Statues* and *Le Tiers-Instruit* are reprinted (with my translations) with permission of Editions Julliard (Paris, France).

Excerpts from Michel Serres and Bruno Latour's *Conversations on Science, Culture, and Time*, translated by R. Lapidus, University of Michigan Press, copyright 1995.

Excerpts from James Gleick's *Chaos: Making a New Science* (New York: Viking Penguin, 1987) are reprinted by permission of Carlisle & Company, LLC, as agent for James Gleick. Copyright 1987 by James Gleick.

Excerpts from Alex Argyros's *A Blessed Rage for Order* are reprinted with permission of The University of Michigan Press, copyright 1992.

Excerpts from Michel Serres's *Genesis*, translated by G. James and J. Nielson, The University of Michigan Press, copyright 1995.

Excerpts from Michel Serres's *Détachement* are reprinted with permission of Les Editions Flammarion (Paris, France), using the English translation by permission of the translators, Geneviève James and Raymond Federman.

Excerpts from Michel Serres's *Les cinq sens* are reprinted (and translated by myself) with permission of Société des Editions Grasset et Fasquelle (Paris, France), copyright 1985.

Also gratefully acknowledged are permissions to reprint excerpts from the following articles:

Thomas P. Weissert, "Dynamical Discourse Theory," in *Time and Society* Vol. 4(1): 111–133. Sage Publications Ltd, copyright 1995 Sage.

Mitchell J. Feigenbaum, "Universal Behavior in Nonlinear Systems," in *Los Alamos Science*, Premier Issue. Reproduced courtesy of *Los Alamos Science*, Los Alamos National Laboratory.

H. W. Brands, "Fractal History," in *Diplomatic History* 16 (Fall 1992), Blackwell Publishers.

Richard T. Gray, "The Dialectic of 'Enscentment': Patrick Süskind's DAS PARFUM as Critical History of Enlightenment Culture." Reprinted by permission of the Modern Language Association of America from *PMLA*, Vol. 108 (1993) (3): 491.

Thomas M. Kavanagh, "Michel Serres, Les cinq sens," in *MLN—Modern Language Notes* 101.4 (September 1986): 937–941. The Johns Hopkins University Press.

Introduction

In the opening pages of *Genesis* (*Genèse*, 1982), Michel Serres offers what amounts to a sort of retrospective view of his earlier writings. Recalling his reliance on traditional concepts of unity, he promises to discourse now instead on multiplicity. What I thus summarize is an oversimplification, by necessity of reductive commentary.

The promised approach is at work in his writings long before *Genesis* and contributed, no doubt, to the acclaim Serres was given, but also to critical assessments of his interpretative strategies, in the wake of the sequential publications of five *Hermès* texts (1969–1980), *La communication*, *L'interférence*, *La traduction*, *La distribution*, and *Le passage du nord-ouest*, as well as *La naissance de la physique* (1977), and *The Parasite* (1980). Something was moving and shifting in these vast epistemological surveys of Western culture and thought. It was not yet a paradigm change, in the sense in which epistemologists today talk about ruptures and dislocations in the sciences. On the surface, Serres had used scientific models, among them most prominently information theory, to uncover parallels with cultural systems and philosophical conceptions and theories. By grounding the sciences into everyday, ordinary life, he hoped to reduce the awesome distance which is perceived to exist between the two and, at the same time, to free the humanities from crushing analytical theories and an empty nominalism. However, these writings are also rich in object lessons which reveal close similarities between new and often disturbing insights gained in the two domains. Significantly, parallels are drawn between the acknowledgment by today's scientists that classical scientific models represent ordered systems that play a rather exceptional, even marginal role in nature, on the one hand, and the acceptance, on the part of philosophers and literary theorists, of the precarious state of all deterministic value systems in postmodern cultural thought, on the other.[1] Serres's outlines for a unified discourse in which these parallels coalesce into singular models—the mythical figure of Hermes, the parasite, the Lucretian clinamen—were and still are the focus for much of the criticism leveled at Serres. He brought a radical twist to the controversy over the degree to which scientific discourse and literary and philosophical theories can coincide without weakening the persuasive force of their respective models.

Serres's epistemological pursuit of a universal discourse used concepts of information theory, which had developed mathematical expressions of probability to structure the relationship of message and white noise in the channels of communication between a sender and a receiver.[2] Of primary concern to this theory is the question how much of any intrusive noise enhances or diminishes the truly informational value of a message as it passes from point of origin to point of destination. Serres's interest focused on these "channels" where information and noise mix into a "new" message, a mixture of anticipated and unanticipated data. Calling them "passages," he investigated them in the most varied forms and through multiple models. They shed new light on a deep-seated yet fluid relationship between science and the humanist domains. As part of the general debate on interdisciplinarity, this object lesson was necessary and timely.

Less obvious at the time, but as important to Serres's place in this debate, was a discursive landslide surreptitiously set in motion in his works. They displayed a growing tendency to include the teaching medium in the object lesson. As proposed by Serres, the notion of fluid passages holding the parallel discourses in close epistemological kinship modeled by Hermes and the parasite begins to waver as their author recognizes—and expresses at the outset of *Genesis*—that the identical discourse he is striving to formulate depends on a concept of unity. For it implies both globalization and stasis, two historical epistemological tendencies he had opposed in his studies as leading to intellectual oppression and stagnation. Serres's hesitation was actually part of a broad movement of reassessments occuring in the sciences as well as the humanities. The era of the 1970s swept the humanities from the modern into the postmodern era, with inevitable reverberations and acrimonious disputes among specialists. On the scientific scene, it saw the arrival of complexity and chaos as valid field models for objective inquiry, with implications well beyond their already established mathematical and theoretical domains. The scientific community, too, experienced consternation and occasional incomprehension on the part of physicists, chemists, and economists who, trained in the classical paradigms, were not willing to share their colleagues' single-minded fascination with turbulence, cloud formations, eddies, random stock market swings, or dripping faucets, and arrythmia of the heart.

Though they will not figure in the investigative part of this study, the last of the *Hermès* texts, *Hermès V: Le passage du nord-ouest*, and *La naissance de la physique* exhibit some of the fluid characteristics of the "passage" in their own textuality, while they deal at the same time with issues that rocked the humanities and the sciences. *The Parasite* holds a special place in this phase, because it models the fluid passage in the notion of the excluded middle (of binary logic) who/that is no longer willing to remain excluded.[3] Aided by the fact that in French the excluded middle is expressed as "the excluded third" (*le tiers exclu*), that is, by a linguistic instance favoring a numerical metaphor, Serres analyzes it from the perspective of a parasitic operator whose intrusive activities set the stage for the "multiple" object of investigation in *Genesis*. This textual fluidity, here at work in the flawless collapse of signified and signifier (the concept of the excluded and the image of a third), is one of the targets of often repeated objections to Serres's epistemological vision. The criticism concerns in particular the question of his continued commitment to rigorous scientific expression. In *Chaos Bound*, N. Katherine Hayles's critique of Serres's theories is based to a large extent on the tendency in his discourse to incorporate "equivocal definitions," "rhetorical moves," and "metaphoric slippage." In particular, she questions the connection between noise (of information theory), the excluded middle, and the "third man" (198–99). This is a valid argument, for the collusion of signifier and signified underscores the risks to the rigor of any rational discourse. To the French reader, however, the appearance of willfulness in the anthropomorphic conversion is attenuated by the lexical configuration of these terms in French. The metaphoric "slippage" may therefore be perceived instead as metaphoric creativeness, in this instance as well as many others that appear in Serres's texts.

By the same token, it is immensely useful for Serres's discourse that in French "le parasite" bears a triple meaning: an organism living on or off another organism, a sycophantic guest who overstays his welcome, and static or white noise. Evolving from the parasitic "excluded third," and thus ineluctably tied to, or identified with, its intrusive and noisy character, the Serresean notion of the "multiple" becomes in *Genesis* a first metaphoric paradigm in a chain of tropes that Serres continues to weave into a series of epistemological writings. Instead of a new set of concepts or a mere realignment of accepted conceptual strategies on which to build his projected discourse, Serres posits what precisely had always been excluded from conceptual frameworks: the non-unitary, loose aggregates, the deviate, the disorderly and monstrous, all that cannot be recuperated into binary deterministic systems or linear thinking.[4] He renames this excluded middle "the multiple" because it comprises a vast, open region in which binary concepts merely account for limit situations and linear systems remain singularities. This "multiple" should not be understood as the multiple of a number. It is non-quantitative, its emergence from the numerical metaphor of the "third" notwithstanding. Sharing the "parasitic" nature of the excluded third, it may be thought of as a "chaotic" number which is contradictory to linear thinking, or as an unchecked multiplication of the "middle" that crowds in on what is normally non-excluded.

Serres presents this intermittently in his texts; that is, almost always his ideas go through a series of images and stories before they are presented as such. The reader needs a sharp mental eye to catch them in their metaphorical stages in order to be receptive to an "odd" or strange idea once it is formulated. The multiple is such an instance. When Serres writes that it "is possibility itself, it is a set of possible things, it may be *the* set of possible things" (*Genesis*, 22), he is implying a dynamical "system" full of chaotic possibilities where time plays a central and constitutive role. A little further, he demonstrates this temporal implication, however, with a spatial image embedded in a text by Honoré de Balzac, *The Unknown Masterpiece*, in which the constitutive action of time merely leads to endless possibilities. The argument seems circular only if one reads into the "merely" a restrictive meaning. But in Serres's text it is a sign of utter, unrestricted, and monstrously liberating exclusion, so utter in its force that it turns the "known" into a creative "unknown." This portrait of the multiple is only the first of many, but it already betrays time's new role, as we shall discover in chapter 1.

Serres's figurative strategy is a bold step, contrary to scientific thought and practices. It is risky for a philosopher and historian of science seemingly to abandon scientific rigor and analytic strategy in favor of an epistemological story that strings together metaphor after metaphor and ends with an image of chaos as a musical Ur-expression. It was a provocation on the part of a scientific mind, and Serres's colleagues in philosophy and science departments who know and esteem the *Hermès* series and *The Parasite* were duly provoked to react in ways that ranged from admiration for a maverick thinker to incredulity, and finally to outright rejection.

In the sense that Serres's discourse now shifts from paradigms of fluid mechanics, passages, and parasitism to tropes of the multiple and the chaotic, I consider *Genesis* to be a new beginning, stylistically speaking. Often described in terms of a shift from a scientific discourse to a more poetic expression, terms that imply regret or criticism, the texts gathered in this study are, nonetheless, best understood as a continuation of Serres's far-reaching reflections in the earlier writings. In these, his epistemo-critical meditations are centered on a philosophy of fluids, understood as an epistemological version of the physical science of fluid mechanics. It is proposed in order to counterbalance traditional thinking analogously represented by laws that govern the mechanics of solids. The overriding metaphorical play, however, is a deep-seated opposition between stasis and mobility, between the solid and the fluid, between repetitive stagnation and inventive creativity. Much later, Serres himself refers to these writings as a "cycle" of epistemological reflections based primarily on concepts of communication theory. He identifies the five-volume series *Hermès* and *The Parasite* as part of this cycle which he considers coming to its completion only with the more recent *Angels: A Modern Myth* (1993).[5]

I believe another cycle or series can now be discerned, among Serres's writings published between 1982 and 1991, where the stylistic shift to the chaotic multiple serves as a point of departure and leads, via a philosophy of "mixed bodies" or mixture (*mélange*), to a new understanding of time. My grouping differs from one presented at the end of the French version of *Angels: A Modern Myth* where Serres's texts are arranged into three categories, with *Genesis* and

Les cinq sens forming part of a section called "Energie et Transformations," and *Detachment*, *Statues*, *Le Tiers-Instruit*, and *The Natural Contract* included in the section "Equilibre et Fondations." Together with a third category, this overall ordering of Serres's writings indicates three related epistemological themes for which the individual works offer their own distinct perspectives. These the present study does not neglect; close readings will establish their validity as independent epistemological issues of postmodern debate.

However, the grouping I am proposing also allows for a global approach to Serres's thought, so that a progressive development becomes visible, from the chaotic multiple to circumstances, to dynamical statues, to a portrait of dynamical cultural systems that *create time* as an operating factor of their inventive drive. The fluid passages of Serres's texts that lead to this insight betray, not a linear development, but a progression based on iteration: Acting as narrative background, the metaphoric chains form a feedback mechanism by which an insight once attained becomes a metaphoric input for the next argument which, in turn, deploys its metaphoric power to produce yet another insight, and so on. To give an example, the multiple as a metaphoric device in *Genesis* already implies the "circumstantial" encountered in *Les cinq sens*, the dynamical stasis that becomes the paradoxical theme of *Statues*, or the new time of invention in *Le Tiers-Instruit*. This text, in turn, celebrates the multiple as its message. The metaphoric nature of the iterative process prevents it from becoming a linear progression and assures an open-ended variability.

At first glance, the six texts gathered in this study do suggest a more or less chronological linearity, but a thorough understanding of the metaphorical iteration allows the reader to see the practical necessity of such an orderly arrangement for the present reading, which does not negate the creative process of the chaotic multiple. Serres's attempts to enlarge the epistemological field of a philosophy of fluids and embrace domains traditionally viewed as of a purely humanistic purview result in a new philosophy of time gleaned from a global reading of the metaphorical process at work in his writings. The present study offers as its objective such a global panorama in a sustained form and, with analogous applications of key concepts of nonlinear dynamics (chaos theory), traces the theme of time as a new epistemological issue. Time is no longer thought of as a parameter adding something to a system *from the outside*. In other words, time is no longer understood as a purely historical force which Serres argues is the source of inherently violent foundations of our epistēmē. Serres's notion of a new time which arises out of nonlinear systemic processes and characterizes his new epistemological approach bears features of a time that plays a "new and unexpected role" in the physical sciences of today (Prigogine and Stengers, 10). Serres's texts present these topics in discursive nuggets, fragments, and sometimes allusive parables whose thematic and conceptual cohesion is not always easily recognizable. In the following chapters, I hope to facilitate this recognition.

It is my belief that such a global insight is a necessary condition for a full understanding of Serres's emerging philosophy of invention. With it, he places himself squarely in the camp of poststructural thinking. His methodological tools are similar to, and at times identical with, deconstructive reasoning. For brevity's sake, the latter may be summarized in Vincent Leitch's formula for deconstructive activity: " Repeat and undermine" (*Deconstructive Criticism*, 177). For the critic, such deconstructive production "is not self-effacing repetition or doubling commentary"; rather, while using these methods, "it produces the imperceptible: the trace of—the supplement, for instance" (Leitch, 178). This deconstructive perspective is focused on textuality and its inherent blind spot. Jonathan Culler's summary of pure Derridean deconstruction goes beyond textuality toward a larger, philosophical formula which then becomes necessarily more general: "Derrida does not do away with such traditional concepts as truth, logic, values, etc., but displaces them in some strategic way" (*On Deconstruction*, 150).

Serres participates in these activities of uncovering the imperceptible trace and displacing traditional concepts so that they become unhinged and free from their traditional logical moorings. However, since our insight of Serres's philosophy of invention is tied to a new understanding of time and will be formulated through a discussion of concepts of nonlinear dynamics as they appear in the Serresean texts, it is also useful to look at the basic difference between deconstructive thought and chaos theory. For example, although "deconstruction shares with chaos theory the desire to breach the boundaries of classical systems by opening them to a new kind of analysis in which information is *created* [emphasis added] rather than conserved" and both "concur in assigning a positive value to chaos, . . . scientists see chaos as the source of order, [while] poststructuralists appropriate it to subvert order" (Hayles, 176). Although Serres displaces traditional concepts, he limits rather than subverts order and assigns to chaos an equally productive source of possibilities.

But a decisive alterity of Serres's thinking resides in his increasing reliance on an epistemological perspective based on dynamical processes. Contrary to deconstructive analysis, the central role that time plays in his perspective, coupled with nonlinearity which the scientific concept of chaos implies, is the source for the notion of the fluid multiple. From this conceptual complex comes our understanding of a truly inventive life. Serres's fervent argument for constant pursuit of invention, and his passionate plea for a pedagogy of invention express a revolt against an ever more persistent homogeneization of our life and an ever increasing leveling of our cultural expressiveness. For Serres, the question of an inventive life is, in the final analysis, an ethical one. The disappearance of variety and the increase of stratification are directly related to linear static thinking. But linear thinking leads to absolute order and exclusion of the other. This, in turn, is the root of evil (*le mal*). Serres equates stasis and linearity with immobility and stagnation. His writings demonstrate a logical connection between complex topology or the transformation of space, dynamics, a new notion of time, the problem of evil, and dynamical invention as a postmodern—epistemocritical—action and response to evil. At the risk of speculating about a definitive term for a work that defies easy descriptions, I advance *invention* as perhaps the singular notion that may be said to traverse all of Serres's writings, under different guises. It is the dominant trope of his discourse on which he concentrates his efforts to present the scientific, philosophical, and literary discourses of Western epistēmē as parallel and, in the final analysis, identical discourses. The present study examines this radical proposition of identity of discourses through a critical analysis of the role time plays in Serres's essays.

I do not intend the present work to be a comprehensive introduction to Serres's writings. Certain texts will not figure here because they record reflective digressions into specific areas of cultural activity (texts on Auguste Comte, Gottfried Leibniz, Émile Zola, Jules Verne, Vittore Carpaccio). Others, such as the studies of the "cycle" based on concepts of communication theory, will play a strictly supportive role. They have been presented in a finely balanced anthology, *Hermes: Literature, Science, Philosophy* (1982), by Josué V. Harari and David F. Bell, who combined selected chapters of the texts involved with an encyclopedic introduction to Serres's work as a whole as it existed at the time. In addition, a postface by Ilya Prigogine and Isabelle Stengers traces implications drawn from these texts for future scientific-philosophical thinking. Many passages in these earlier works lead directly into the arguments set forth in the texts discussed in the present study. For instance, on the backcover of *Hermès IV. La distribution*, Serres relates order to disorder as a rare island archipelago to the vastness of the ocean. If the goal is to describe order, then systems and structures must be outlined and analyzed. On the other hand, acquiring knowledge of disorder means that one has "to leave the islands and plunge into the sea" teeming with multiple manifestations of chaos. But if one wants "to eliminate the latter, only a *functional description* is available" [emphasis added]. Such descriptions must integrate localized experiences into global structures, formulas, and laws, if understanding is the desired objective, and always imply an outside observer. This has also been his strategy, Serres writes, and he adds that "there has never been a *nova scienza*, except perhaps in one instance of which I speak in my next book, the *gaya scienza* promised by a return from order to disorder" [author's translation] (*Hermès IV*, 12). The next book is *La Naissance de la physique dans le texte de Lucrèce*, closely followed by *Hermès V. Le passage du nord-ouest*, perhaps also *The Parasite*. Together they prepare a blueprint for a new science of local knowledge which eschews globalizing theory.

However, in her commentary on "Serres's commitment to local knowledge" (*Chaos Bound*, 197), N. Katherine Hayles faults the philosopher for "giv[ing] way to the synthetic impulse to make the parasite into a universal concept in its own right" (198). She goes on to show that this is an example of his equivocation which "works (if it works) only by assimilating obvious differences to apparent similarities. And yet the explicit intention in creating it was to liberate the noise of empirical differences from the oppression of idealized similarities" (Hayles, 198–99). Serres's retrospective characterization of his earlier works as relying too much on the concept of unity springs from a similar critical assessment. The promise of *Genesis* to become a book on multiplicity recommits him to the "plunge" into empirical localities and to take a further step toward the *gaya scienza*. It does not abandon all "functional description" of the new science. But it is also the portrait of a "happy science," because the plunge into chaos transforms the discourse into a participant of the turbulent processes it describes. A functional description requires an outside observer, but the *gaya scienza* beckons the outsider to join the fray. Serres's retrospective musings about previous writings modulate into an observation that the passages he pursues so ardently in many domains of our epistēmē have come into play in his own epistemological survey. His parasitic third now moves towards and implies chaos, or the *multiple*, in his own discourse qua epistemological discourse. In a certain sense then, *Genesis* is a provocative gesture that gives encouragement and approval to the erratic behavior of Serres's parasitically multiplying model. The distinction between universal similarities and empirical differences becomes meaningless or, rather, unknown.

Other texts that followed the publication of *Genesis* and share its joyful pursuit of localities reinforce a general critical observation that Serres abandoned accepted scientific discourse in favor of a more "poetic" expression. Much more than a critique of a possible stylistic shift, these comments imply regret over a manifest shift from theoretical, global paradigms to an amalgam of local knowledge that can never come to the point, as one critic expressed it, or form a coherent theory other than being "poetic." In a 1987 article, Bruno Latour who later collaborated with Serres on a text to enlighten the reader (*Conversations on Science, Culture, and Time*) describes the dilemma, so evident in Hayles's critique, of understanding what Serres is all about, in terms that put his thinking into the contemporary context of French intellectual life where he has his roots. For Latour, Serres is an empiricist in the sense that he talks about things without having recourse to the idealist's trick of talking about *our knowledge* of things. "To my knowledge, this is a rare attempt, in philosophy, to see things from the point of view of the *known*, not of the knowing" (Latour, 89). Latour calls Serres a "queer sort" of an empiricist, because for a French person to be one is to be "totally un-French . . . in his philosophical tradition" (89). Although Latour is entirely supportive of Serres's empiricist revolt against postwar phenomenologist trends in France, his remarks indirectly demonstrate the difficulties the Serresean reader encounters while trying to see the *known*, if s/he happens to have no knowledge about it. For it must be said that Serres's argument progressing from "the point of view of the known," that is, from empirical evidence, presupposes nonetheless a certain degree of knowing about, or related to, the known. Even in *Les cinq sens* where sensate experiences appear to be the most successful at bracketing the knowing and approaching a sheer empirical known, Serres himself laments the presence of language as the inevitably interfering instrument of cognitive knowing.

Reading Serres is therefore a reading challenge. We have to learn how to read provocatively, an activity I describe in later contexts as a dynamical reading.[6] It is a necessary evolutionary step in the art of postmodern reading, if we wish to comprehend fully what is at stake when an epistemological enterprise such as Serres's passes from hermeneutic reflections on a vast tradition of scientific, philosophical, mythological, literary, and artistic expressions to a discourse that reflects *topologically* the results and consequences of this encyclopedic program on an internalized level. Theoreticians of complexity describe such feedback loops as sets of interrelated elements emerging out of other or "lower" sets, and variously call them emergent phenomena, nested hierarchies, and dissipative systems. In the pages that follow, the notion of emergence with its concomitant self-organizing phenomena will serve as a determining concept and as a recursive guarantor of non-closure for our dynamical reading. I shall frequently refer to dissipative systems which, in contrast to systems that conserve their energy, experience loss of energy as a result of their dynamics. If a system does not absorb new energy from outside sources, it is fated to end in motionlessness. If it interacts with outside sources of energy, these will become part of the system's dynamics. But since not all initial conditions of such acquired parts can ever be known completely, the resulting dynamical system displays unexpected tendencies which alter and thus renew the system. This self-organizing, inventive activity allows us to discern a paradigm shift in the Serresean discourse, one that emerges out of a global reading of the texts that preceded those we are concerned with in this study. The former foreshadow and treat the concepts at issue here but on a different hierarchical level and from an observer's perspective. The following chapters describe the self-organizing phenomena that emerge, in turn, from this paradigm shift.

This shift is not a sudden rupture or about-face; rather, it represents a slide from the author/observer's "functional description" to a level where outside and inside have lost all distinction. It is the "passing" from the "north-west passage" (*Hermès V*) to the "passing in the middle-place" where the *excluded middle* learns to become an instructed-middle (*Le Tiers-Instruit*). It is a shift of perspective from the "problem of space" to the problem of time. Hermes, "the philosopher of plural spaces," must topologically solve apparently incongruent juxtapositions of epistemic domains, because "the conception of space is crucial to Serres's epistemology" (Harari and Bell, xxxiii). On the other hand, the problem of time, obliquely broached in *Genesis* as a "pure multiplicity" (21), is treated in *Statues* and *Le Tiers-Instruit* in such a way that appears what dynamicists call the "timing of space."

In an interview published in *The French Review*, Serres acknowledges the shift the present study deals with. He speaks of an evolution in his writings and describes its five facets; among them is a new tendency on his part to "orient" himself more toward beauty than scientific "demonstration," because "beauty is an inherent part of the labor of writing" (James, 788). Lest one think, however, that Serres has abandoned the scientific track, we must keep in mind that science, too, has altered its discourse. As an example of discursive interferences and *mélange*, I cite H. -O. Peitgen, perhaps the best known interpreter and illustrator of Benoit Mandelbrot's fractal geometry, who titled his work on this topic, "The Beauty of Fractals." The result is a stunning compilation of computer-generated mathematical works of art preceded by an introduction which is accessible to the nonscientific reader, while other chapters deal with related, advanced mathematical and technical topics. The analogy between an aesthetic concept such as beauty and scientific-theoretical notations, here only alluded to, becomes more evident in the following chapters where fractals will play a role as isomorphic models for our reading of the Serresean discourse.

Since we traditionally interpret time either through its historical manifestations and cyclical progressions, or as the natural clock in and of physical systems, time will often appear here in these contexts. Chapters 1 and 2 of the present study contain many references to historical time, the natural clock, and their effects upon our epistemological conscious, in order to underscore by contrast the other "new" time emerging from beneath the guise of tropes. Many of the references to linear time, as it manifests itself in historical consciousness, will be associated with René Girard's theory of the scapegoat mechanism. The latter derives from Girard's theory of mimetic desire which explains desire for a beloved object (or person) as being in reality the imitation of another's desire for the same object (*Deceit, Desire, and the Novel* 1965). This triangular structure of mimetic desire collapses into a bipolar attraction when the other's desire itself becomes the focus of the self's desire. The other appears at one and the same time as model and rival; he represents all that the self lacks. This Hegelian interpretation of a universal human condition, which pits the self against the other the more the other is admired and imitated, allows Girard to posit that mimesis leads to rivalry and conflict. Since it is a universal trait, it leads to violence of all against all and to a universal threat of extinction by self-inflicted generalized violence.

After extensive ethnological and anthropological research, and by logical inference, Girard arrived at the theory of the scapegoat mechanism which explains how a solution to this murderous situation is provided (*Violence and the Sacred* 1977). A particularly random but spectacular conflict that leaves one of the rivals victorious can arouse such fascination in others that they imitate the violent act (as always, mimetic desire is involved) and fall upon the same victim. Violence by all directed against all changes into violence-in-unison against one. Concentrated upon one scapegoat, violence is safely deflected from all others. A necessary condition for the formation of society is thereby achieved, at the cost of one randomly chosen excluded victim who bears forever the onus of evil as well as the mark of a savior. The scapegoat is considered as both a malevolent and benevolent outsider. This monumental exclusion is constantly repeated to guarantee the evolution of the collective community. Society is thus founded on an act of violence by exclusion, while history is the chain of repetitive imitations of this act.

Where Girard sees the formation of the sacred in this linear time of founding exclusions, Serres sees the same juxtaposition of violence and the sacred, but he also recognizes the essence of evil as such at the core of the repetitive gesture of exclusion. To overcome evil, Serres proposes an inventive rather than an imitative effort; to achieve this, the linearity of historical time has to be replaced by a new time. The texts which figure at the center of each of the following chapters are read primarily as discursive simulations of such a new notion of time. As a sign of evolving emergent phenomena through which fateful historical foundations are loosened, chapter 3 will celebrate a certain detachment finally attained, as well as an understanding of time gained in empirical encounters through the sensate.

The present study will make frequent reference to nonlinear dynamics, a theory which is presently at the forefront of a new conceptualization of time. Serres himself applies some aspects of this modeling strategy to summarize his implications for the sensate experience, in the later part of *Les cinq sens*. Nonlinear dynamical systems theory (also known as chaos theory) is "the qualitative study of unstable aperiodic behavior in deterministic nonlinear dynamical systems" (Kellert, 2). Nonlinearity is a precise mathematical definition for certain types of equations; in conceptual terms, it means that a minute change in the initial makeup of a nonlinear system can have disproportionately large and unpredictable consequences leading to future instability in the system. (In a linear stable system, small variations produce small effects, large changes proportionately large ones.) The concepts involved in this scientific theory may be used to develop a new epistemological understanding, as Alex Argyros proposes in *A Blessed Rage for Order*:

> [T]he most robust and complex things in the universe, including human beings and their theories, works of art, and social structures, are best understood neither as Platonic essences nor as random processes, but as chaotic systems. Chaos might be in a position to render outdated the old distinctions between necessity and chance, . . . replacing them with a dynamical theory that encompasses both local unpredictability and global order. Insofar as chaos theory contributes to reversing the dominant contemporary understanding of the relation between order and freedom, . . . it promises a new ontological and epistemological paradigm. (Argyros, 6–7)

In Serres's essays, the loss of repetition inherent in the nonlinearity of chaotic systems is of outmost importance for the formation of inventive thought.

Chapters 4 and 5 of the present study will make extensive use of simulations of the nonlinear dynamical model, by conceptual analogy. *Statues* and *Le Tiers-Instruit* teem with dense metaphorical strings of apologues which make at times for difficult reading. The use of the dynamical model prepares conditions for a global reading pattern facilitating the understanding of "localized" tropes without sacrificing an integral comprehension of the texts' ultimate vision of time and invention. Chapter 6 deals with *The Natural Contract*, a text that on first sight does not seem to fit in with the writings I am attempting to present here as a coherent group of texts on time. However, the legal concept of prescription, used by Serres as one of the tropes in *Le Tiers-Instruit* to reveal a relationship between time as a "natural" phenomenon and the question of right as an expression of the human contractual conscious, is globalized in *The Natural Contract*. "Prescription" foregrounds all arguments presented in this text which may therefore be read as a demonstration and concretization of time at work in a postmodern epistemologically renewed Lebenswelt, in the most literal sense of this expression.

An explanation is necessary for the role *Rome: The Book of Foundations* with its essentially Girardian perspective of historical foundations plays in the present study. The "epistemology of human relations" (Harari and Bell, xxxix), modeled in *The Parasite* on the excluded third as a parasitic operator, is enlarged in *Rome*, and becomes an epistemological archeology of repetitive layers of historical and cultural founding. More importantly, two aspects of this work set it logically apart from the texts that are the object of the present study. First, the question of death as the absolute "radical Evil" (*Hermès V*, 130) is not only embedded in its interpretations of repetitive exclusions, like a destructive force counterbalancing cultural and historical creativity, but also leads to cataclysmic or apocalyptic implications. In the six works that are the objects of our reading, on the contrary, death evolves into a matrix for creative forces, while the problem of evil is stated as epistemological immobility versus inventive freedom and is metaphorized into a question of suffering and tolerance.

Second, and closely related to the first aspect, *Rome* is the ultimate expression of the irreversible linear arrow of action by which Serres characterizes the evolution of any parasitic operation. It has a predictable destination that is summarized in *The Parasite* as the "end of the *Odyssey*, amidst the corpses" (339). In the following chapters, I translate such an outcome as a fixed-point attractor. In striking contrast, *Les cinq sens* pleads for action based on a "Ulyssean meandering" (*la randonnée*), with no fixed destination to cut short its inventive course; therefore, without corpses which are the markings of death triumphant. Exemplary for the texts on time, it shows that Serresean discourse now tends more toward non-closure which, in terms of nonlinear dynamics, means a promise of global knowledge with local unpredictability. In chapters 4 and 5, I translate this with the help of conceptual models related to "strange attractors." Supported by Serres's own references to these attractors (*Les cinq sens*, 287), the models I am borrowing from complex dynamics will illustrate my argument that the texts here studied are more readily understood if we read them like *dissipative* open systems. They reveal time from an entirely new perspective; they reveal that today, parallel to the physical sciences, the evolution of epistemology has reached a point where "on every scale self-organization, complexity, and time play a new and unexpected role" (Prigogine and Stengers, 10).

The story of *Rome*, however, is modeled like a conservative system with a predictable equilibrium: repetitive cycles of violent foundations. Its images of historical time's destructive force are intensified versions of the parasite's noisy and clamorous activities. Already earlier, Serres describes *The Parasite* in its own postscript as a "book on history, a book of Evil," as if infested by a "horrible insect." It returns with ever greater fury in *Rome*, where the author recognizes, from the very outset, that Evil reigns on its pages which describe a "terrain of terror" (epigraph for page 1) built on closure—

as long as they speak *only* history. For I must now relate a story of deliverance from historical time's stranglehold on *Rome: The Book of Foundations*; it reveals the kinship of its last pages with Serres's reflections on the new time which I identify in the present study. The passage I am referring to recounts the seventh day of rest in a long story of the creation of Rome, written by the Roman historian Livy, but read and copied by Serres in his own fashion (276–82). Having given his account of the history of terrorizing and murderous foundations that brought forth Rome, our author describes how, on an imaginary sun-day on the height of each summer season, a Roman ritual was performed on the Fields of Mars alongside the Tiber (279). For Serres, the annual ritual symbolizes the loosening of death's grip on Rome, because all activity ceased on that day; the people enjoyed rest and peace. "History is erased" (280). In the symbolism of the ritual, Serres discovers the true foundation of history: Livy writes that ceremoniously harvested wheat was thrown into the seasonally receding waters of the Tiber; he adds that these sacred grains would have germinated and, with other flotsam and sediment of the river, slowly built up a terrain like a land-fill—the terrain of Rome. Here, Serres discovers a different foundation: historical time takes a rest and is effaced, but history is founded—on physical reality, on the natural world of plants and sand, stones and water (282). Clearly, the "conception of space" is still very much at the center of Serres's epistemological discourse. Yet it cannot entirely overshadow something new in these pages: a new time that is not purely historical or defined solely by human activity, but is also founded in physical reality and on non-closure or dissipative emergence. This founding relationship will be Serres's epistemological concern in the writings which are discussed in the following chapters; it will be expressed through a new conception of time.

Serres often repeats that producing "books made of other books" or books written about other authors is perhaps popular and the preferred methodology of literary critics and theoreticians, but that it has nothing to do with creativity, invention and new thought. Aimed at his contemporaries in academic circles who often answer with disdain or distrust, this challenge carries a measure of the distance Serres put between himself and conventional academic customs, for at least the last fifteen years. His appeal to a wider public has grown in inverse proportion to the academic and especially the scientific cold shoulder.

In my attempt to present a reading strategy for certain of Serres's writings that would reflect the inventive thought of the Serresean text as a whole, I am aware of two pitfalls built into this endeavor. The first is that the union of reading and inventive writing may not succeed in the ideal form I envisioned when I set out to undertake the task. After all, the ideal reading I am proposing here is colored by personal preferences and consistently influenced by subjective parameters that may not or cannot easily be transposed to objective or ideal levels. Second, and I readily admit that this is the much more gnawing or personally disturbing of the two, the enterprise I am about to undertake goes against Serres's very urging not to make books about other books. Though it may indeed be a valid critical enterprise, it may fall short, nonetheless, of the ideal cognitive endeavor that so profoundly characterizes the very text my enterprise tries to present and above all to elucidate. Hovering from the outset like a shadow over this study, therefore, is the onus of seemingly ignoring *for itself* the very message it tries to convey and carry from the Serresean text to the reader.

By the same token, however, this "primer" on Serresean reading tries to practice what the philosopher advises to do. Serres is adamant on characterizing his own praxis of poiesis as one of finding and not of searching; finding, which implies an element of randomness, is the action of a truly creative person (*Le Tiers-Instruit*, 163). Those who write books on books are, as we all know, researchers; they re-search what somebody else before them has already found; the researcher only reuses what he or she already knows. He who finds, on the other hand, is a troubadour in Serres's eyes, an assignation whose linguistic roots are related to finding. The inventive creator is a troubadour, whether he lived in the thirteenth century or in our contemporary time. Serres hopes to gather those who find, because in them lies the promise of *invention*, a keyword that appears with increasing frequency in his texts, until it becomes the leitmotif for *Le Tiers-Instruit* in the context of reflections on a new knowledge. Serresean invention is isomorphic to what in dynamical systems theory is referred to as dissipative behavior of complex nonlinear systems, that is, its own creativity changes along its own inventive activity. It is of no small interest to this reader that the logical transition Serres establishes between finding (*trouver*) and inventing (*inventer*) has its linguistic counterpart in German where *finden* (to find) becomes *erfinden* (to invent) by the mere addition of a prefix. In general usage, the prefix *er* for German verbs is defined as expressing "achievement of the aim set by the action [of a given verb]" (*New Cassell's German Dictionary*). According to this formula, Serres's logical transformation is supported by a linguistic combination—albeit in another language—that sees in invention (*er-finden*) the achievement of the aim set by any activity of finding (*finden*).

I claim the double notion of finding and inventing for my defense against the stigma of being an epigonic writer. Yes, the present study obviously is a writing on a particular set of books. But it is also a testimony to a continuous finding while in the process of being written. It gives testimony to a pure finding that far outweighs the research that was the catalyst at its inception. The impact of the Serresean texts discussed in the following chapters rests not so much on an understanding of what I was searching for, but above all in constant surprises of the mind, sudden insight unexpectedly stumbled upon, and bifurcations that took me far away from the text still under my reading eye. My reading Serres was an ever-increasing interaction with my growing understanding. As it evolved over time, it rapidly became an activity that displayed, and still does, characteristics which are analogous to those characterizing natural systems that are in what Ilya Prigogine calls "far-from-equilibrium states." They remain in open-ended evolution.

With this inventive reading, dissipation becomes a positive term metaphorizing its association with distractions, digressions, and idle playfulness into states that literally spring surprises and never point to closure: I *am finding*, in the Serresean sense. But I wish to add a transitive connection for my reader: I found the bifurcating power of metaphors where before I had searched so arduously for them in poetry (chapters on *Detachment* and *Les cinq sens*); I found an ontological justification for ecological concerns which goes far beyond any politicized environmental movement (chapter on *The Natural Contract*); I found Johannes Kepler, Sir Isaac Newton, and Lazare Carnot, Jules-Henri Poincaré and Benoit Mandelbrot, classical mechanics, thermodynamics, dynamical systems theory, all in a glorious, exhilarating rediscovery of half-forgotten school subjects, with the intellectual pleasure and satisfaction of new discovery added, finally understanding because I simply found it there (the *ci-gît* of *Les cinq sens*). Whatever I gathered in these findings went far beyond the (re)searching process and eclipsed mere efforts to fill the gaps of scientific illiteracy which get wider and deeper the more the humanities person in me demands my personal and professional attention and time. I found phase-space concepts and strange attractors, buzzwords of chaos theory, translated into a truly ontological understanding of the human phenomenon (chapter on *Statues*). I found the making of the human miracle in a "birthing" process that evolves over time into the "instructed-third" person, for the full epistemological meaning of which the French reader has the advantage of knowing that the traditional "third" is both an excluded middle and the spirit of some divine triad (chapter on *Le Tiers-Instruit*).

And invention came in the wake of finding.

And if I can prove it here and there, now and then, between the covers of this book, then I know this enterprise is not a research or a copy, but a creative undertaking resembling liberating fiction.

In the following, quotes from sources other than English are taken from published translations where such exist. Where none exist or did not exist at the time of the writing of this text, I use my own translations.

1

Time Promised: Reading *Genèse*

> Et moi je parcourais, cherchant qui me réponde,
> De degrés en degrés cette Babel du monde.
> —Victor Hugo, "La Pente de la rêverie"

In the five texts of what is now often referred to as the *Hermès* series, Michel Serres investigates the nature of "passages" which he identifies as links between the physical and the human sciences. Borrowed from concepts of information theory and portrayed allegorically by Hermes, the god who communicates, the theoretical model Serres develops is most often circumscribed by the old French expression "noise".[1] In *Hermès I: La Communication*, this generic term expresses the "set of interference phenomena that become obstacles to communication" (40). Thoroughly detailed in *Hermès IV: La Distribution*, the "theory of noise" is then concretized in *The Parasite* with the triune allegorical figure of the parasite who puts the theory into practice. The observations on parasitic behavior complete and legitimize the theory. Noise itself has become a phenomenon, and Hermes and the parasite its messengers and heralds.

For the English-speaking reader, in particular, it is important to comprehend the use of the complex term "noise" in the French context of Serres's writings; it is necessary for a complete understanding of the "multiple" as a first model for the new concept of time. The present chapter tries to elucidate this complex connection for which a lexical contrast in Serres's works may serve as an initial illustration. In *Feux et signaux de brume. Zola* (1975), scientific concepts of thermodynamics provide the base for a unified discourse on philosophy, history, and the writings of Émile Zola. A "festival of fire" (100) is the engine that drives this text and leads to the typical Serresean notions of noise and multiplicity. But here, whether molecular chaos, biological species, or historically moving events are being discussed, noise is always expressed by "bruit," which unequivocally means noise in French, while multiplicity is used in the sense of large numbers, "grand nombre" or "nombre élevé" (176–83). Serres develops these themes by reading Zola's *Les Rougon-Macquart* in which "the object of the text is consistently some pot-boiler" recounting crushing uprisings, ruinous family relations, and self-destructive individual actions. Time is no longer counted out by the cyclical return of clock hours; rather, it is steered by a "material law." Time is the irreversible force of entropy; it is the "time of the cauldron," time "produced in the coke oven" (181).

"Noise," however, slowly replaces "bruit" as a key notion in Serres's texts. Bruit becomes part of "noise" which the author revives from old French where it meant disorderly furor as well as noise. Similarly, "multiplicity" becomes more frequently "the multiple"; and the "large number" changes imperceptibly into an infinite set or, as I call it in the introduction, "a chaotic number." The shift from bruit to noise, from a clearly understood modern word to a term more or less effaced by historical time is symptomatic for an evolution of Serres's materialistic notion of time. From the irreversible arrow of entropy, only individually stemmed by neguentropic action, the discussion moves toward a concept of time as an integral part of dynamical systems, where entropy and neguentropy resolve into the interplay of global and local events. This is traceable over many of Serres's texts and, although clearly discernable in *Genesis*, fully implied only in later texts, as we will see in the following chapters.

Built into the very concept of "noise" as a set of interference phenomena, and in "noise" as the parasite who triples as an abusive guest, a parasitic organism, and/or static noise, is the overriding notion of the excluded middle or third (*le tiers exclu*) without which the entire logical structure of Western thought is unthinkable. Josué V. Harari and David F. Bell give a concise summary of *The Parasite* and very rightly suggest that Serres "could just as easily have called it "The Demon" or "The Third Man," because the exclusion itself becomes dominant (*Hermes*, xxvi–xxviii). We may see this dominance in the way Serres's readers have recuperated the parasite as a workable model for theoretical discussions in various fields of the humanities. In an interesting comparison of Serres's and Gilles Deleuze's respective approaches to communication through comedy and humor, Peter Hertz-Ohmes uses Serres's Hermes model to illustrate a conservative, circulatory system of exchange at work in comedy, while the parasite plays a mischievous operator and, in the final analysis, provocative participator within. For this critic, therefore, Serres's models are far more conservative and "orderly" than Deleuze's "crack-up" of order, or the "betrayal" of Hermes which he models on the "Aeon, the eternal instant that momentarily countermands the reign of Hermes" (Hertz-Ohmes, 246). Richard Braverman uses Serres's parasite and his elaboration of a theory of the third man to analyze the libertine-courtier figure of the seventeenth-century theater and of English Restoration drama, in particular. Yet, while discussing Serres's presentation of Molière's *Dom Juan* from the perspective of the theory of the parasite, Braverman regrets that Serres never asks "what power accrues to the parasite" from its/his disruptive behavior, and that the libertine-parasite becomes "the harbinger of a new social function in theory only" (Braverman, 76). In other words, the concept of "noise" should be structured and systematized, the critic seems to suggest, so that it can be effectively applied to sociohistorical analyses.

As one more example of the usefulness of Serres's parasite, I cite William Paulson who very purposefully uses the parasitic model, or rather its thermodynamic aspect, to illustrate a "cascading" movement of economic operators in Balzac's *Le Cousin Pons*. He closes his article with an invitation to other Balzac researchers to avail themselves of Serres's concepts like "good parasites" (Paulson, 1985). The use of the parasite as a "preferred" excluded one and as a marketable model is, of course, very apt but shows the same feedback tendencies displayed by the scapegoat mechanism that René Girard explored in *Violence and the Sacred*. The excluded middle is a safeguard against the violence of disorder and chaos. But its mechanism is effective only as long as a certain degree of violence in the form of expulsion is tolerated and even assumed to be beneficial. Analogous to Girardian sacrificial violence, this controlled and acceptable violence becomes the unitary concept of the Serresean parasite.

Genesis is an attempt to give prime billing to the exclusion, but *without* its conceptual dominance or "accrued power," and *before* it is legitimized as the excluded third by a logic based on the principle of identity, that is, on unitary and rational concepts. If a rational binary recuperation of the parasite is to be avoided—which was Serres's goal all along and is explicitly exemplified in *La naissance de la physique*—the alternative has to be considered. Disorder, chaos, the brouhaha of both physical nature and the clamor of human relations have to be discovered, uncovered, and accepted as valid fields of contemplation without squeezing them into a theoretical straight jacket.[2]

The theory of noise, perhaps theory as such, is called into question in the opening pages of *Genesis* (4). We may assume that time, too, becomes once again a problematic notion. Reflective commentary on texts selected from French literature are Serres's descriptive tools to broach the theme of time, the consequences for human beings when time is understood in historical context only, and its close relationship with the problem of evil. The author begins to *listen* to "noise" and invites his reader to do the same. Through an active hearing, a "new object" for philosophy is thus proposed: the multiple. "Noise" as a complex of obstacles to communication becomes background noise for all information, a sort of Ur-noise of genesis proportions, comparable to formless matter out of which Aristotelian forms arise (7). Easing "noise" out of the phenomenological connection, Serres "transplants" it to a prior, founding state or condition. Now posited pre-phenomenologically, he associates "noise" with yawning chaos and the empty void. This arch- or Ur-noise (*le bruit de fond*) is not merely background noise, which would be a partial rendition of its significance and therefore misleading. It was inevitable and constituted a logical progression that "noise" should give Serres cause to reflect on his previous work and admit that he had "never attempted to conceive of the multiple as such, directly, without ever allowing unification to come to its aid" (4).

The task is daunting but not impossible. The "novel object" Serres is proposing for philosophy is isomorphic to the new geometry the mathematician Benoit Mandelbrot presents in coherent fashion in *The Fractal Geometry of*

Nature, for the same compelling reasons that drive Serres to his chaotic task.[3] Fractal geometry constructs models which simulate forms occurring in nature that are so complex that Euclidean geometry cannot adequately represent them. "At its heart is the idea of a fractional dimension. Whereas Euclidean shapes can be represented very well by the familiar integer dimensions of a Cartesian space, irregular forms cannot. The corrugations that mark their surfaces give them, in effect, an added fraction of a dimension" (Hayles, 164). A fractional dimension is difficult to visualize, as difficult as it is to produce abstract patterns of the many irregular forms and processes of nature. Fractal forms display scale-independent symmetry when examined in ever finer detail. It means that a process of self-similarity occurs with every added fractional dimension observed. The resultant open-ended complexity of the model simulates irregular shapes of naturally occurring phenomena. "Thus one speaks of *approaching* [emphasis is Hayles's] the natural form, rather than representing it exactly" (Hayles 166).

In Serresean discourse, the process of "approaching" is translated by the notion of *implication* with which Serres replaces explication. We shall see that the new philosophy of time cannot be explained and analyzed, it can only be implied. It is for this reason that we will eventually come back in more detail to fractals and their function in chaos theory. For now we need to understand that, compared with Euclidean geometry which offers ideal, orderly models of a complex world, "nature exhibits not simply a higher degree but an *altogether different level* [emphasis added] of complexity" (Mandelbrot, 1). The mathematician is calling attention to a qualitative shift, not merely to a quantitative increase of complexity that one might observe with the help of technologically more refined research tools. Serres is equally insistent about finding new implications because, as a model, the parasite tends to be exploited at higher and higher levels of complexity but is incapable of representing a *different* level, one "that challenges us to study those forms that Euclid leaves aside as being 'formless,' to investigate the morphology of the 'amorphous'" (Mandelbrot, 1). Himself a geometer by training, Serres expresses his novel task in almost identical terms: "[The multiple] is not an aggregate, it is not discrete. It is a bit viscous perhaps" (*Genesis*, 4–5). The multiple is the metaphorical vehicle Serres now chooses to guide his reader through chaos, that is, to have him confront the complexity of the most common elements of his world.

Our first task is to investigate how "noise" as the "expanded" or chaotic parasite becomes the multiple. Three interconnecting levels, philosophical, compositional, and linguistic, offer an answer and explain why it is so crucial for the reader to operate the same transposition from the "third" to the chaotic multiple. The first level has already been alluded to. In its superb metamorphosis as a parasite, "noise" is still a phenomenological event, thus subject to conceptual recuperation. As such, it is prey to the dialectics of human relations and potentially a mere addition to or a new loop in the circular logic of the unitary, rationalist mind. *The Parasite* is the overwhelming demonstration of this process by which the excluded third insinuates him/it-self into a given system only to become, in turn, the system per se. Inevitably, the excluded/now-included is exposed to or threatened by a larger and mightier parasite who will repeat exactly the same steps. Harari and Bell correctly name Serres's study of this process "an epistemology of human relations" (*Hermes* xxxix) and emphatically stress that the problem of human relations "might be the most suggestive aspect of *The Parasite*" (*Hermes* xxvii).

The "multiple as such" proposed as a discursive model in *Genesis* must be freed from these phenomenological shackles if it is to circumscribe the founding of a new epistemology that will no longer be part of the Girardian circle of exclusionary violence. Therefore Serres states unequivocally that he is searching for means to do without theoretical tools and crutches which lead to "syntheses, whereby we shove multiplicities under unities" (4). The multiple can only be discovered if we grasp it without resorting to theoretical proofs, critical analysis and debate, and combinatory rules (syntheses) for what is variable and perhaps oppositional. In other words, Serres proposes abandoning the methods of traditional philosophy in order to arrive at the "new object" that a new philosophy should and must concern itself with. For, just as scientists have come to recognize that Newtonian classical mechanics presents only a fringe model of the real world which is disorderly, postmodern epistemologists recognize that classical philosophy privileges logical, ideal situations and relegates illogical, ill-fitting phenomena to pathological and mythological domains.

For traditional repetitive thinking, it is indeed a foolhardy strategy to throw away—seemingly—the tried and true tools of rational thought empowered by a long and celebrated history. But Serres has no intention of throwing out the child with the bath water. He claims the right to be an *inventive* philosopher, not merely a spokesperson for binary thinking in the yes/no mode or all/nothing dichotomies, but also and in particular for that other neglected and grey area of

> knowing a bit, a meager amount, enough, quite a bit; there are various undulations, even in the hardest and most advanced sciences. We are confronted with or absorbed by the multiple, where more than three quarters of our knowledge and actions are concerned. (5)

Instead of being an exception, the multiple is the most common of situations; yet it has always been set aside as too messy to handle, by the sciences and philosophy alike. Mandelbrot refers to the messiness of nature as "divergence syndromes":

> Some quantity that is commonly *expected* to be positive and finite turns out either to be infinite or to vanish. . . . Such misbehavior looks most bizarre

> and even terrifying [to a scientist], but a careful reexamination shows it to be quite acceptable . . . , as long as one is willing to use *new methods of thought.* [emphasis added] (*The Fractal Geometry of Nature*, 19)

Serres's inventive philosophy forges such new methods. By putting the multiple squarely on the front burner of his enterprise, "the object of this book" (title of the introductory chapter of *Genesis*) is new to philosophy but common and ordinary to everyday experience. Eschewing the phenomenological crutch for his new philosophical approach, Serres is left with what is there, "before" objects are perceived as phenomena by the human mind. He is left—but it is everything—with "noise" as the chaotic implication of what he had so ardently and logically pursued through the excluded third in the figure of the parasite, itself a bifurcation of the Hermes-principle.

Privileging one of the three aspects of the parasite, *Genesis* is a book on noise as Ur-noise. Listening to its multiple din is the first step to knowledge: "Hearing is a model of understanding" (7). The model of traditional philosophy is seeing; the word idea derives from the Greek verb *idein* meaning to see. But seeing is intermittent (7), always already beginning in the phenomenological mode, and can be halted at will by closing one's eyes. Hearing is continuous and never rests; in the deepest silence, "noise" is there and can be heard, if we are willing to open ourselves to the cacophony of the Ur-noise.[4] The first level for understanding the passage from "noise" to the multiple or chaos is thus established. It is entirely philosophical and implies, not a rupture with traditional philosophy, but an enabling strategy for going back to a more "original" understanding of the objects of our world and, by association, of human relations. Among the *Hermès* texts, *La Distribution*, in particular, deals with the question of "noise" as a manifestation of chaotic ensembles. In fact, Serres suggests that we understand "distribution" in a pre-combinatory sense as an *éparpillement*, total dispersion or scattering (14). To grasp these turbulent ensembles that surround the orderly instances we call our "epistēmē," we have no other means but "functional descriptions" which always fall short of a complete grasping. For better understanding of what is utterly incomprehensible because it is chaotic, *La Distribution*, too, proposes a switch from seeing to hearing according to a process that is explained through cybernetic concepts.

More radically in *Genesis*, Serres urges the same switch from seeing to hearing but in a metaphorical language that posits noise as fundamental to the effectiveness of true hearing, comparable to the importance of Plato's *eidos* to classical philosophy. "Noise" is the undifferentiated, uncategorized, indiscrete aggregate that may be called "the multiple." Therefore, Serres ends his introduction on "the object of this book" with a general statement that is both a precursory announcement of the core of his argument and a final summary of his retrospective reflection: "The multiple had been thought (*pensé*) perhaps, but it hadn't been sounded (*entendu*)" (8). The full impact of this statement can only be discerned, however, if the participle *entendu* is transposed into English in its full meaning of both hearing *and* understanding, so that the statement needs to be repeated: "The multiple had been thought (*pensé*) perhaps," but had not been understood (*entendu*).

The second level concerning the question of "noise" is compositional by which I mean the contents as well as the physical composition of *Genesis*. In order to do so, we must step out of the text itself and consider the writing that surrounds it: an *hors-texte* in the Derridean sense of this expression (see Jacques Derrida, *La Dissémination: Hors livre*). *Genesis* begins with a preface called "A Short Tall Tale," while the French original, *Genèse*, carries in addition a postscript on its back cover.

At first glance, the back cover blurb seems to follow the traditional French practice of the *prières d'insérer*, originally loose inserts containing pertinent information on a new book and created by its author, then slipped into copies intended for its reviewers. But a second look allows for new insight into Serres's strategy for noise. Instead of functioning as a capsule for the book's content, the back cover information explodes it: the reader learns that the title is not really its title, and that the title it ends up with is also incorrect, because *genèse* should be plural to reflect "abundant possibilities" for beginnings. The title was meant to be "Noise" and was in fact "Noise" until Serres was prevailed upon to change it to *Genèse*. One may speculate that this was an editorial decision influenced by the fact that "noise" is really a word few French would find catchy because of its outmoded use, while *genèse* is provocative and authoritative at the same time, at once rivaling and deferring to the Bible. But more significant for the subject matter and its attempted pre-phenomenological mode is the fact that the title change becomes a postscripted issue, on the back cover. For, by recounting the evolution of his title, Serres legitimizes the editorial reading, inadvertently or perhaps deliberately, into a correct reading practice for the discussion taking place within the text: "Noise" was to be there, metaphorically, becoming and fading, chaotic and ordering, as loud furor and Ur-silence, but always and above all pre-phenomenological. "Noise" as title, however, would have posited the very word in its full phenomenological authority; titular "noise" would have run the risk of becoming a phenomenological model similar to the parasite.

In "The Double Session," Jacques Derrida used a well known line by Stéphane Mallarmé about the art of forgetting titles because they speak too loudly, to introduce his deconstructive reading of a Mallarmean text (*Dissemination*, 177–180).[5] For him, Mallarmé's deliberate suspension of the title is much more than a simple decapitation, so that the rest of the page "resists the authority and presumption of the title, the plumbline and aplomb of the heading" (178). The play between the whiteness of the page and the black traces the poet will impose thereon is marked first and foremost by the initial active forgetfulness of the title, understood as a fundamental strategy: "Mallarmé recommends, then, that silence be imposed on the title but also that one draw upon it as upon the resources of a germinal or seminal blank" (179). The title is liberated from the logical constraints of a conceptual entity and recognized as that chaotic or blank source out of which spring the phenomena of our conceptual world. Mallarmé's forgotten title becomes Serres's title "noise" decapitated by his editorial readers but then suspended by the author on the back cover. This gesture of a quasi reinstatement avoids reducing "noise" to one of the many phenomena the text is concerned with, and points instead to its seminal role as the Ur-noise that underlies our phenomenological perceptions and *backs* them up. The metaphoric scenes of *Genesis* are thus supported by a back-cover "noise" that has been forgotten as a phenomenon, since as such it spoke too loudly and too parasitically overconfident. But from its hors textuality of the back cover, it can now be heard as the arch-noise or Ur-noise deeply embedded in the arguments of the text.[6] By insisting on a change of title, Serres's first readers participated in the philosophical reservations with which the author enters into the argument proper of *Genesis*. By recounting the discussion he had had with them, Serres in fact repeats his cautious reserve vis-à-vis traditional philosophical habits, through the reaction of others, and prepares his reader for the "new" philosophy by *hearing* "noise."

The discussion of the second, compositional level is not complete unless a word is said about the other *hors texte*, a sort of preface called "A Short Tall Tale." (The original French text calls it "Conte court.") Printed in italics throughout, it resembles passages in *Genesis* where Serres quotes other authors. These citations are identified only by their italicized imprint; no references are given.[7] With the "conte court," it appears that Serres is quoting himself; the chaotic multiple begins its work: The preface is the text, and the rest (the actual text proper of *Genesis*) a commentary on "itself" in a feedback loop manoeuver. All this is only allusive and never explicit, but utterly equivocal in the sense Hayles uses this term when discussing Serres's work.

A Short Tall Tale

> As I was sailing along that summer, under a dazzling sky, and drifting lazily in the wind and sun, I found myself, one fine morning, in the green and stagnant waters of the Sargasso Sea, at a mysterious spot where thousands of tiny sparks, all shapes and all colors, were glimmering crazily in the early morning light. Bearing off, I was dumbfounded to see an area almost two hundred and fifty acres square entirely populated by dancing bottles. There were countless little vessels, and each one no doubt bore its message; each had its freight and each had its buoyant little roll, ballasted with seawrack and rockery; each carried its hope and its despair. The coiling winds had compelled them all there, from far and near, from a thousand different

> quadrants. Their constant and perilous collisions made for an acute and cacophonic carillon, and this noise mounted heavenward, wafted to the horizon, it filled all space with giddy ecstasy.
>
> The following night, a wide sargasso put me in danger of shipwreck. I had just about foundered. Swiftly I made a raft of some of the bottles, they worked well as floats and bladders, and thus did I make my way back to Bordeaux. (*Genesis*, 1)

Barely filling two thirds of page 1, it is separated from the rest of the text by a blank page. In the manner of a fable, it preempts all that follows. But as with all that is simple, humble, and suspended without a title (after all, "Conte court" refers merely to genre and length of the text, not to its content), what follows is necessary for its understanding in the sense of hearing, because we seers of ideas have forgotten how to hear and listen. The narrator is sailing the North Atlantic, leisurely giving himself and his craft over to the haphazard forces of wind and ocean current. My geography text tells me that the Sargasso Sea shifts between latitudes 20° and 40° and longitudes 35° and 75°. A particular kind of marine algae called "sargassum" grows here, often in such dense accumulations that ocean ships avoid this region. These tentacled plants hinder navigation and in the past posed a grave threat to smaller vessels with less sophisticated instruments. For the reader, the mysterious aura hanging over this region echoes tales of the Bermuda Triangle. All this is of no interest to the narrator who is trapped in these "green and stagnant" surroundings. What catches his attention, however, are a million little bursts of sunlight refracted from acres and acres of bottles tangled in the algae tentacles and bobbing in the water. Of course, the reader knows that these are the proverbial bottles tossed into the sea, each holding a message, each with a bit of knowledge, gathered up as haphazardly by the gulf stream currents as the narrator's sailboat drifting with the forces of the elements. But neither sight nor message have an ultimate impact on the solitary sailor. Both sunlight (nature) and knowledge (the messages of human relations, that is, culture) have broken up into innumerable discrete units; multiplicities have replaced the unitary mold.

The story's dreamlike atmosphere is punctuated with "measurable" references: a *thousand* little bursts of light; a *hundred* hectares of watery expanse; the winds blowing from a *thousand* directions; an *incalculable number* of bottles, which is an infinite number useful in mathematics even if incalculable; the *number one* in the form of each bottle; a measure of weight ("each bottle had its freight") and of motion ("each had . . . its buoyant little roll"); and finally, bytes of information ("each one no doubt bore its message, . . . its hope and its despair"). All this is secondary because it is phenomenological. A deeper, more elemental and indistinct force was there that engulfed all discreetness, all distinctive and measurable features: The encounter of these discrete units, banging and clanking together, amplified a millionfold, produced "an acute and cacophonic carillon."

There is an overwhelming din, clamor, or undifferentiated "noise" that fills the story, so to speak, for nothing more need be said: "This noise mounted heavenward, wafted to the horizon, it filled all space with giddy ecstasy."

The story ends here, for the sailor has heard chaos, furor, the multiple; he heard the pre-phenomenological origin and understood it. The final paragraph of four lines tells another story. A tale of mortal danger and potential shipwreck, it demonstrates that hearing and understanding chaotic "noise" without the aid of concepts is a new learning: After a day of listening to the siren song of chaos, night fall confronts the sailor with the deadly force of nature. One large sargassum alga has ensnared his craft so badly that it is in danger of capsizing; he will perish at sea. But out of "noise" comes rescue in the form of bottles lashed together to make a raft. The chaotic multiple endangers his life but releases also enough self-organizing information. Man creates a unitary model from the myriad elements surging up from chaos, builds his life-saving support, and returns to his logical world. (Bordeaux is part of Serres's Gascony origins.)

The preface figures as a correct projection of arguments which are presented but also weakened in the main text of *Genesis*. More specifically, it anticipates the first chapter in which Serres reads Balzac's "The Unknown Masterpiece" (*Le Chef-d'oeuvre inconnu*) in a way that grasps the "unknown" as the chaotic multiple and finds the "known" in the aesthetic manifestation of a painting born out of creative chaos. The preface projects this epistemological ground work, on the level of "noise." The first and principal paragraph relates the dreamlike drifting into "noise," a passing that will later be described as a necessary descent into chaos, where the philosopher must "speak to the obscure shadows that cannot take shape, and rake over the ashes" (126). The cacophonic noise that fills the world in drunken folly is the fuzzy shadowiness of the prelogical netherworld metaphorized into noise. The second, less prominent paragraph demonstrates a practice in hearing "noise," because the narrator understands, that is, he correctly uses the objects that human work has fashioned (the bottles) and nature has herded together (with the help of seaweed). Neither the messages inside the bottles, carriers of human relations with their inherent pitfalls of violence, rivalry, and clamor, nor the algae as carriers of nature's cruel force, but the bottles as pure objects *and* the sargassum weeds as pure objects, mysteriously uniting human technology and nature's dynamism, have a meaningful self-organizing message for the mariner.[8] The bottles' din and random racket produce the idea of ordering their noisy collisions among the green lianas into the lifesaving geometric form of a raft. Having heard (understood) "noise," the narrator is reborn and surges forth out of the waters, like a "new Aphrodite" (*Hermès II. L'Interférence*, 190), cradled in the security of his world, as she is on a seashell.

The third level concerning the question of "noise" constitutes a linguistic or stylistic perspective. I have already hinted at the fact that "noise" plays a significant role in *Genesis*, in spite of the ban of its use as title. If Serres is to maintain its pre-phenomenological role as the fuzzy "somewhat" by which we are to hear the chaotic multiple that makes up our world, "noise" must above all be deprived of any conceptual substance that may still be attached to its name. In this endeavor, Serres finds a collaborator in Balzac and his story. In order to hear "noise" without immediately conceptualizing it into a thing or experience called noise, Serres abandons the noun in favor of an adjectival construct, *noiseuse*. The neologism is borrowed from Balzac, although the latter had deleted it from his story in the course of correcting subsequent versions. It is the feminine form of the nonexistent *noiseux*, with the suffix *eux* producing the adjectival form of "noise." Introduced as the dominant metaphor in the first chapter of *Genesis*, *noiseux* appears almost exclusively in the Balzacian form *la belle noiseuse*, loosely translated as the beautiful noisemaker, the beautiful clamorous (one), and/or the beautiful cantankerous (one). *La belle noiseuse* becomes Serres's generative metaphorical figure for modeling a reading of "The Unknown Masterpiece" that will demonstrate a story on chaos, and ultimately of a new notion of time. But for now, the shift from "noise" to *noiseuse* must hold our attention. For it parallels the relegation of its titular position to a back-cover blurb. The shift from noun to adjective represents much more than a stylistic gesture affected by the author. In the case of *la belle noiseuse*, it is only the beginning, albeit a spectacular one, of a long and continuous Serresean practice which favors a language where adjectives, adverbs, and prepositions play more than supporting roles for nouns and verbs, the stalwarts of syntactical constructions for Western languages. They become semantic entities in their own right and thereby offer inventive possibilities for the multiple. Although the French language, much more so than the English, facilitates the substantivizing of any part of speech, Serres's preference for adjectives, particularly in the case of *la belle noiseuse*, is based, not on a desire to substantivize, but on an explicit intention to hear, show, and *legitimize* the nonsubstantival word in its own essential nature: as an *accidental*.

Once again, a comparison with new scientific thinking is helpful in underscoring the potential impact of Serres's linguistic maneuver. Mandelbrot calls his investigations of the morphology of the "amorphous" a new geometry and modifies it with the neologism "fractal." In the past when faced with the amorphous, "mathematicians have disdained this challenge . . . and have increasingly chosen to flee from nature by devising theories unrelated to anything we can see or feel" (*The Fractal Geometry of Nature*, 1). Similarly, classical disdain for adjectives and other secondary parts of speech is rooted in philosophical discourse and goes back to Aristotelian thought. The hierarchical division into substances and accidentals was translated under the influence of the scholastics into grammatical structures mirroring the same hierarchical dichotomy. It effectively reinforced the overriding authority of the noun which names everything in imitation of divine creativeness. We must add that in French "noun" and "name" are expressed by the same word, *nom* (m.), and that the classical, authoritative voice in matters of French grammar was the so-called Port-Royal Grammar (seventeenth century) which reaffirmed the Aristotelian distinction, for ideological and theological reasons.

Analogous to the decision not to grant it titular authority, Serres posits "noise" in the accidental form *noiseuse* for the hearing-understanding of the multiple. He chooses the adjectival form in full awareness of its tangential status; it has no substance and, more importantly, does not really exist in the French language and never did, even when "noise" itself was a current term. In its almost meaningless, nonsubstantial, nonauthoritative setting, therefore, *la belle noiseuse* reflects linguistically the fluctuating, fuzzy, viscous somewhat that is pre-phenomenological and formless. Lexically, *la belle noiseuse* not only prevents "noise" from revealing itself as a clear shape or form, but also helps establish an amorphous reading of the multiple.

Hermès IV. La Distribution proposes a new epistemological understanding of the relationship between nature and culture attained through a functional description of chaos. *Genesis* continues this work but endeavors to go beyond the portrait of chaotic signs that betray the otherwise ineffable presence of disorder in our ordered world. Serres now lets go of the raft of concepts and its promise of a home port. He remains in the midst of the cacophonic din of noise in order to "hear" it *inconceivably*. This precarious undertaking goes hand in hand with the evolution of the parasitic "third" toward chaotic multiplicity. Both are preconditions for the long journey to a new understanding of time, as we shall discover.

The introductory chapter of *Genesis* is a scientifically chaotic and philosophically pre-phenomenological attempt to present a very physical and philosophical reality: chaos as a generative multiple. In the body of the text, this is pursued under various metaphors of which *la belle noiseuse* is the most prominent and perhaps best known. One could argue, therefore, that the objectives of this text are a metaphorical refinement of previous writings. This is indeed the case; yet, the means to deliver evolving Serresean thinking have changed qualitatively, as the discussion on the lexical shift from "noise" to *noiseuse* hoped to elucidate. Merely assuming that more recent writings by Serres are studies on a stylistic shift from scientific language to a more poetic expression, is a superficial observation that misses the essence of his encyclopedic critique of Western epistemology and his dynamic quest for new approaches to contemporary issues in the cultural and historical realms.

In the *The French Review* interview of 1986 (referred to in the introduction), Serres confirms what *Genesis* demonstrates in its textual composition, namely, that "logic is inseparable from style" (James, 793). This Flaubertian formula, slightly bent to accommodate the philosopher's argument, summarizes an interview that above all tries to expose a new aesthetic which cannot simply be truncated or dismissed as poetic. Based on the Greek roots of the word poet in the verb *poiein*, to make or fabricate, Serres's argument reconnects aesthetics to philosophy to ethics to science, in the mode of interacting passages (James, 790). These are familiar to the reader of the *Hermès* series and *The Parasite*. They establish conditions in which the epistemological domains remain fluent and dissipative, that is, open and receptive to systemic changes. Mandelbrot says the same with his new geometry of nature requiring "new methods of thought." His call echoes the words of Ilya Prigogine and Isabelle Stengers who speak of "new tools of thought" (*Order out of Chaos*, 203). Newness or revolution in thought are keywords in these discussions which, Serres stresses, denote, at the same time, nothing new or uncommon. *Mélange*, mixture, and the multiple discern, not a qualitative change of reality, but an essentially different mode by which we humans must relearn how to look at reality. Prigogine and Stengers emphasize repeatedly in their study that one of the major differences between Newtonian classical science and twentieth-century scientific thought comprises a different view of the role played by the human observer of nature. For Newtonian thinking the observer remains outside the observed system to safeguard the objectivity of the observation. Modern science, however, has recognized that he or she is part of the observation without necessarily subjectivizing it.

This new perspective governing scientific approaches is expressed to a high degree in Serresean writing and intertwined with his aesthetics. Serres connects beauty with philosophy and science; in turn, his philosophy is the quest for an expression that does justice to the interfering, connecting, and crossing passages of epistemological domains. The question of (poetic) style is therefore of utmost importance for *Genesis*, but not in the sense of a critical literary assessment. Rather, it is a question of a bittersweet, even ironic choice for the author:

> I was condemned to invent a new vocabulary which would have complicated the situation [of presenting a new philosophy] even more, so little by little I resolved to use more and more natural, everyday language. But the moment you refine language as much and as well as possible, you create a style. (*Conversations on Science, Culture, and Time*, 71–72)

As it develops in *Genesis*, Serres's style is not a mere fanciful play by an epistemologist and historian of science, with his beloved language. The few remarks on Serres's stylistic shift suffice to show that borrowing *noiseuse* from Balzac is more than an admiring salute to a great French author. Furthermore, the primary intent of *Genesis* is not to introduce the reader to literary perspectives of the chaotic workings of Hermes and the parasite, with the inevitably concomitant abandonment of scientific rigor. As an historian of science, Serres finds that certain literary works betray an inherent and succinct understanding of chaos and the multiple as something other than disorder or monstrous abnormalities, long before nonlinear, chaotic and nonintegrable processes in nature became objects of serious discussion among scientists. Therefore, "the multiple is the object of this book and history is its goal" (*Genesis*, 7), to which we add time as the most problematic of the text's objectives.

La belle noiseuse is the first model of *Genesis* by which the author makes his reader hear the multiple. In its most obvious association, it marks Serres's reading of Balzac's story about three artists, the young French painter Poussin, the mature Flemish painter Porbus, and a fictional character, the old master Frenhofer. Central to the story is a painting which is nothing but a confused assemblage of colors and meaningless brush strokes, the result of years of minute, maniacal refinements by the old master. Around this painting, Serres uncovers a *temporal process* from young to old: Poussin's artistic representation with linear, bold strokes becomes a mastery of multiple shadings in the work of Porbus, and ends in Frenhofer's portrait of confusion. He likens it to a genealogical tree that traces the chaotic multiple through an "order of representation" the artists have dedicated their lives to. But on the level of reality, that is, according to the "order of being" embodied in three women who each give meaning to their respective artistic endeavors, this hierarchy seems reversed, for it passes from destroyed, chaotic beauty to the nude perfection of a live woman (12). The genealogy becomes confusing for Serres who can no longer determine the direction and meaning of time (both direction and meaning are expressed in French by *le sens*) as either a creative movement toward a perfect masterpiece or as the entropic reality of the human condition of aging and destroyed beauty. Both movements intertwine. Two genealogies emerge from the story and seem to cross along the same tree (12). The succession from root to crown is as unpredictable as the movement from crown to root. Time remains the "unknown" at the heart of "The Unknown Masterpiece."

But Serres takes a second look and begins to hear. In later versions of his story, Balzac had eliminated the original appellation *ma belle noiseuse* by which the old master affectionately recalled the beloved and chaste model. He had thereby carefully covered up a disorderly element in his story that, if left undeleted, would have introduced a logical inconsistency into the exposition of the aesthetic principle the story contemplates.[9] After all, the Balzacian narrative hinges on the absolute necessity to demonstrate convincingly that Frenhofer sees pure Beauty in the (now chaotic) portrait of his beloved Catherine Lescault, a masterwork ten years in the making and consuming all his energies. The appellation *ma belle noiseuse* clouds her "good name" and mars her chaste beauty. Balzac corrects this static noise in the message by deleting it from later versions of his story.

I have elsewhere described the strategy behind Serres's gesture of reinstating *la belle noiseuse*.[10] Undoing the orderly process of the Balzacian text unveils an underlying, pre-phenomenological aesthetic possibility, as a necessary condition for beauty to emerge. From this perspective, Serres's reading of the story circumvents Balzac's clean-up gesture and instead joins the old master in seeing monstrous disorder as beauty. Only from this understanding arises the recognition that Balzac does speak of chaotic and dissipative systems—just as surely as the great nineteenth-century mathematician Henri Poincaré who was fascinated, and often terrified, by "monstrous" nonintegrable mathematical equations, or as Prigogine and Stengers who suggest the creative thrust of dissipative, far-from-equilibrium systems in physics and chemistry.

La belle noiseuse is a metaphor that holds a special fascination for Serres, as a *deleted expression.* Its excluded status gives it the aura of a parasite, a sufficient property for setting any Serresean passage on the multiple into fluctuating motion. In an interconnecting series of frequently allusive references (12–15), "noise" is recognized as vexation and quarrel in modern French, while modern English retains the other part of the old term, noise. Metaphorically speaking, anger and quarrel are separated from noise and din by water, as France is separated from Britain by the English Channel and (French) Quebec by the St. Lawrence River from its southern English-speaking shore (the New England region). Furthermore, there is an ocean between Serres's French roots and home and the English-speaking New World where he found a second (professorial) home. (It is also anecdotically useful to recall that he served for a time as an officer in the French navy.) Banal as they may seem, these considerations color the way Serres works *la belle noiseuse* into the core of his argument with which he intends to present "time as a pure multiplicity" (6). *La belle noiseuse* of the Balzacian canvas is now to be found in the vastness of the ocean, the watery expanse dividing noise and quarrel, "in the midst of the dividing waters" (13). The original French text speaks of *eaux de partage*, which conveys both division and sharing of the meaning of noise in the two languages.

But what we term banal circumstances which bring the sea into this train of thought becomes eminently reasonable when we continue reading with Serres: The old painter's beloved is named Catherine Lescault, a name Serres equates with the river Escaut (the initial *L* being the French article "the"). Its source lies in France (Département Aisne), but its course runs mainly through Belgium where it is called by its Flemish name "Scheldt." The beloved is therefore Catherine of the Waters, says Serres, the waters that divide (*partage*) and share (*partage*) the word "noise." What for so long was divided into two languages is brought together in the strange expression *la belle noiseuse* and then suppressed by Balzac for reasons of narrative clarity and logic. But Serres argues that Balzac eliminated the linguistic expression only because the chaotic masterpiece that old Frenhofer finally consents to unveil before the eyes of his two disciples is already an aesthetic attempt to state the multiple without concepts, and without a nominal representation. In other words, the novelist did not need to add a linguistic sign which would have been redundant. By resurrecting the suppressed surname in the context of a chaotic multiple, Serres is able to connect it—without concepts as well—to a region that straddles both the metaphysical and the physical realms (13–14), and thereby opens this rich metaphor to a philosophical and epistemological reading.

The in-between-ness of being both noise (English) and quarrel (French) lies in the vastness and depth of the sea, in the watery name attached to *la belle noiseuse*. The image of the sea associated with the model must be understood both literally, in the sense of the biographical allusions to Serres's personal history, and metaphorically, in a sense that cuts across various experiences of our cultural life. Serres accepts the often stated etymological connections between "noise," nausea, and nautical (13), although *Littré*, a leading French dictionary known for its careful documentation of associative and etymological derivations, questions the connection on the ground that it accounts only for the part that expresses discord and annoyance, not for the other standing for noise. The etymological controversy has no meaning for Serres, because as a mariner he has experienced the deep silence of the high seas as a primal "tranquil" noise, and conversely the steady, deep, and perpetual brouhaha of the agitated waters as a primal, "vehement" furor (13). The Ur-noise is even called the "ocean-noise" (*bruit de mer*), for it "never ceases; it is limitless, continuous, unending, unchanging" (13), like the ocean, or at least like the cultural concept we have of the ocean as a limitless quantity or entity. Victor Hugo uses the sea in such a metaphysical sense quite frequently in his work, as a mirror image or approximate expression of infinity in our finite world. It is not by accident that Hugolian poetry plays an important role at the moment when Serres's reading of Balzac's story comes to fruition.

The ocean-"noise" is therefore not the combined sound produced by wave motion and wind turbulence, nor the physical violence of nature's raw forces. Rather, it is that other noise one hears (rarely today, alas!) when one finds oneself surrounded by utter silence. Likewise, it is the awe-inspiring potential of brute and hostile furor, felt when one crosses the open, seemingly infinite expanse of the high seas. It is a sort of cosmic primal noise and furor that is never a phenomenon but "lies [just] under the cuttings of all phenomena" (14). *La belle noiseuse* is not one of these phenomenological cutouts; nor is she "noise," but beauty emerging from the in-between waters. I must emphasize once more for the non-French reader, how closely associated by sound, too, "noise" [*nwaz*] is with "noyer" [*nwae*] meaning to drown, inundate, submerge in a fluid medium, so that the association between *la belle noiseuse* and water or fluidity does not only come from Balzac's story, but also from their homophonous relationship enhancing the connection noise-nausea-nautical. However, this is still a metaphor; the question remains how this links up with multiplicity and time.

La belle noiseuse of the waters is fluidity, but she is also Frenhofer's painting. Serres seeks to combine the representational figure with the sea. He thinks first of the Greek sea god Proteus (14–15) to whom mythology attributes two qualities in particular: the art of prophecy and the ability to assume many forms, animate and inanimate. The two traits are linked, for Proteus did not willingly share his knowledge. One had to catch him in one of his forms and do so quickly enough before he slipped back into the formlessness of the waters. Only under one of his assumed forms would he prophesy and thus transfer knowledge. Serres acknowledges the epistemological wisdom on display in this story, for it points to a multiplicity of localized forms cut out from a global Ur-noise. But he also sees its blind spot, because it says nothing about the passage between chaotic "noise" and form: "Who is Proteus when he is no longer water and not yet a panther or a boar?" (14). As always, Serres is interested in the "passage" because there he will find the only pre-phenomenological possibility to understand time. He returns to "The Unknown Masterpiece" where *la belle noiseuse* is not a form but in-between-ness, Proteus-in-passage between ordered form and formless Ur-noise. Of course, Balzac's story does not refer in any of its narrative figures to Greek mythology. The detour of Serres's argument via Proteus permits the reader, however, to see with him in *la belle noiseuse* another mythical figure of Antiquity, Aphrodite, or Venus, and to see her with Proteus-like attributes. For what else can Catherine Lescault's portrait be but that of Venus or Aphrodite? Venus is the paradigm of "beauty of the waters," which we forget too often, because what we admire when we look at Venus are her beauty, grace, harmony of perfect form, and human greatness bordering on the divine and expressed through artistic perfection.

And now Serres performs the most decisive transformation of *Genesis*, for which the digression through the myth of Poseidon's shepherd prepares the stage. Venus rising from the foam of the agitated sea is Catherine the beloved born of the waters of the Escaut. The analogy is based on the element of beauty shared by both (*la belle*). But Balzac originally makes his old demented painter call her "la belle *noiseuse*," which is Venus the cantankerous and noisy one, because the master is the only one who really understands the turbulence of his portrait which does violence to any notion of logic or order. By perfecting his creation with unending additions of color and forms, Frenhofer has achieved "beautiful" chaos on his canvas. The portrait of the woman is a process of drowning (*noyer*) in an agitated sea of incoherent color blotches and meaningless lines. Yet he sees her. So does Balzac who seems to have found a new way to interpret the surname antiquity gave Venus: Anadyomene, the one in the process of rising from the watery foam. But he is careful to attribute such a disorderly understanding of Venus anadyomene to the creeping dementia of the old master.

In a way, Serres catches Balzac peeking at this chaotic truth and then trying to recover what he has seen by eliminating the cantankerous surname. Our author thinks he understands what Balzac and his characters are doing. If we look at Frenhofer's masterpiece and search for anything recognizable, we are like

Poussin and Porbus, "imbeciles" who look for certainty or for phenomena that cannot be doubted (15). Instead of reading into the story a Balzacian attempt at a generalized formula for a nineteenth-century aesthetic principle, as many literary critics do, Serres sees a process. It is the slow evolution from the straight, Euclidean line in young Poussin's work, to a double, trembling, imprecise, fluctuating outline of forms in Porbus's mature painting, to a chaotic multiplicity of drowning, meaningless "forms" created by Frenhofer. The latter's infinite additions make his masterpiece unknowable, and therefore inherently unknown (16). The Serresean transformation has changed the genealogical tree of artistic representation with its static points of reference *into a process with unpredictable future outcomes*. The story models a dynamical process which directly implicates time. We shall later return to this relationship.

For Serres, the artistic progression of the story is also an epistemological process: From the certainty of a linear idea put down on canvas in Poussin's youthful self-assurance, the process goes to Porbus's masterly expression of doubt and bifurcating possibilities. It culminates in the uncontrollable, endless doubting of Frenhofer who is never satisfied with his creation but adds continuously, here some strokes and there endless dabs of color. All his certainties have long ago disappeared in an amalgam of unceasing doubts, arriving at a sort of open-ended, dissipative knowing, which is not the normal rational process of acquiring knowledge. His two disciples finally obtain the privilege of viewing their master's work but, seeing only confusion and chaotic (in)completeness, treat him like a fool who has produced non-sense. Serres, in turn, is somewhat harsh with Balzac for calling the old painter demented. After all, it is through the novelist's story that we recognize the negative aspects of a purely rational pursuit: Whereas Frenhofer's creative process is the epitome of ecstatic doubt and imprecision, rational thought is helpless in accounting for the apparently nonsensical, unless it has a handy pronouncement on a genius gone mad, unless it declares the process a failure by reason of incoherent and disorderly behavior.

Sustained by his inspiration, the artist simply multiplies the open-ended possibilities of perfect beauty. Balzac's insane and genial idea to make the old master not only *see la belle noiseuse* in the jumble of chaotic dots and lines on his canvas, but above all *understand* her in the *process* of bifurcating possibilities, is read by Serres as an aesthetic model that explains the nature of time (17). I shall refer to this process as a dynamical system in later chapters. For late twentieth-century readers who wish to read dynamically, the fascination with Balzac's story does not come from its exposition of a masterpiece as a static work of art that observers try to decipher. Serres mocks Poussin and Porbus for engaging in precisely such a useless undertaking. By attempting to trace out the shape of a woman as a figure of certainty beneath the multitude of colored blotches and fluid forms, they are in effect freezing into one moment the endless process of creative possibilities. They are short-circuiting *time as a dynamic process* which goes from a unitary formula to the richness of the possible. This process "is not potential, it is the very reverse of power, rather it is capaciousness" (22). Time is pure chaotic multiplicity—this is what Serres sees in the unknown masterpiece. Once it is known, that is, once a particular shape is clearly recognized among all the possible combinations of form and arrangements of color, that singular possibility will freeze into a certainty, a phenomenon, a known work, Venus. But all other possibilities, all other multiple choices have to be abandoned. This is the price we pay when we admire Venus-Aphrodite as she appears on the canvas of Giorgione or Botticelli, or as she was created by Apelles, the fourth-century B.C. Greek painter, or Venus of Cyrene and Venus of Knidos.

Or, on the contrary, Venus conjured up by Balzac-Frenhofer-Serres. As *la belle noiseuse*, she is of the waters, amorphous; and yet, she is already rising from them, because something-already-heard/recognized surges from the formlessness of "noise": her beauty. This Venus-Aphrodite is known only as a chaotic painting of amorphous content; yet she is a masterpiece because the master sees beauty on the canvas. If we act like Poussin and Porbus, we are now asking: Is she or is she not on the canvas? Is Venus born of the foamy waters or still unformed like Proteus in the womb of the limitless sea? The question bars all possibilities, except one answer: the binary choice of yes/no and either/or.

> We always see Venus without the sea; or the sea without Venus, we never see physics emerging, anadyomene, from metaphysics. Formed phenomenal information gets free from the chaotic background noise, the knowable and the known are born from that unknown. (18)

The Serresean parasite has been transformed into Aphrodite Anadyomene; she is the excluded third in the ever-open process of being included.

Serres is frequently criticized for his lack of analytical rigor in the way his discourse posits a basic identity between philosophical, scientific, and humanistic (literary) discourses, an identity that goes well beyond commonly recognized parallels in their discursive strategies. His reading of *la belle noiseuse* brings us close to an implied understanding of such an identity. But as with all paradigm shifts, which entail a good measure of rethinking the very nature of explanations, the Balzac-Serres story of process is anything but an analytical account of epistemological identity. Moreover, "explanations under two different paradigms are not just dissimilar; they are incommensurable" (Hayles, 169). Since postmodern thought accepts the notion of parallel discourses in the domains of our epistēmē, but is still cautious about claims for an actual identity of their conceptual makeup, we have to overcome our own reading (explanatory) resistance and pass the threshold of the incommensurably different paradigm.

Genesis does not make this step easy for the reader because a long tradition of depicting the chaotic and disorderly as the essentially feminine aspect of our epistēmē is here uncritically accepted and adopted by Serres for his allegorical presentation of a radically other way of "hearing." In her seminal study of chaos, Hayles is quite emphatic when she writes of scientists who, even in their application of paradigmatic shifts characteristic of chaos theory, remain under the influence of such culturally conditioned thinking (172–74). Since our reading of Serres proceeds towards a nonlinear dynamical mode of understanding, it is interesting to note that the feminine "other" is viewed not only from the same double perspective in *Genesis* as in the research of many chaologists, namely, as creative openness of chaos, but also as a threat of the unfamiliar and uncontrollable. The double name of *la belle noiseuse* expresses this Janus-like approach, in every respect. Although other factors in *Genesis* contribute to its failure to exploit to the fullest a good metaphoric frame that exhibits the new temporality, it is the woman as other who ultimately derails Serres's argument. She is the object of exclusion in the Balzacian story, both as the fascinating and chaste beauty who lures the old master into her exclusion and by implication causes his dementia, and as creative chaos permanently barred from rational sense. Only when Serres detaches his discourse from such Girardian exclusions does the new temporality come to full fruition in his writings. And yet, *la belle noiseuse* is the first paradigm and pivot of Serres's reading of the new notion of time and we have to return to her.

For Serres, *la belle noiseuse* is the *passage* from the pre-phenomenological primal soup to our phenomenologically ordered world. She is the *processing* of all possibilities, not their sum or reservoir. Since she is neither chaos nor order, she is what dynamicists call "a phase transition," that fuzzy state when a system is at the threshold of a phase change. This is usually an event marked by turbulence.[11] Therefore Balzac's *belle noiseuse* becomes Serres's "turbulent Venus" which may be the correct, late twentieth-century transformation of her surname Anadyomene. Turbulence is an as yet undecided openness to all future directions of a system; we may call it "possibilities unlimited." Frenhofer's turbulent Venus is an immensely profound representation of the possible, the multiple. Serres's double process of unveiling the inherently unknowable portrait and reinstating chaotic formlessness as a beautiful and meaningful possibility in *la belle noiseuse* enables him to grasp time as an operator in a "system" of initial doubt that bifurcates toward a profusion of possibles. He admires "the old fool" because, after all his efforts already expended, the aging master keeps saying: "Yet, still I have doubts" (17). Serres sees in him the opposite of the rational mind. Although Cartesian rationalism is based on the premise of fundamental doubt, the Cartesian universe is constructed on its systematic elimination, in a linear process toward certainty. Frenhofer, too, starts with a doubt, but he allows it to double, then redouble and bifurcate until it swells and sprouts to the multiple of possibilities, a true portrait of chaotic dimensions (17). The first of these processes we call "rationalism," the other, "dementia." Serres, however, prefers to call the latter "pure multiplicity," which is his way of putting time before us in the turbulent and noisy portrait of *la belle noiseuse*—without the help of concepts.

For Serres, chaotic fullness of possibilities is at the same time a blankness, emptiness, whiteness. He calls it "the naked multiple" (18) and means a metaphysical nakedness, analogous to the "naked truth." The multiple, stripped of all conceptual veils by which we might shroud and immediately constrain it to a unitary form, must remain naked, smooth, colorless, as well as full of all possible colors. It must remain noisy and clamorous, cantankerous and quarrelsome, while at the same time be void of any conceptualized message; it must remain silent in the midst of its din. Serres shows us how to read the noise and furor, the chaotic and the madness in Balzac's story. But does it also show the silence, the absence, the nothing that global possibility should also be if it is to remain immune to conceptual manipulation?

Serres finds the answer to this question in the figure of Gillette, the girl who loves Poussin. In the wake of youthful rashness and as a kind of dare, the painter had proposed that she pose for Frenhofer, because her beauty would certainly rival the vaunted perfection of Catherine Lescault. Poussin and Porbus are convinced that their master would thereby find a resolution to the endless corrections he works into his portrait. But once they have arrived at Frenhofer's studio and his masterpiece is unveiled, the men forget Gillette who stands nude, abandoned, silent in a corner, waiting for the sitting to begin. As always, Frenhofer is enraptured and in deep contemplation of his beloved. As for Poussin and Porbus, they are confused and remain preoccupied with the enormous effort to make sense of a senseless representation. Finally, they concentrate on one corner of the canvas where they discover a trace of something recognizable. A "foot looms there, alive and exquisite, . . . a strange fluctuation midst the fury and the sound of the *belle noiseuse*, drowned in the chaos of multiplicities" (36).

Serres, however, does not forget Gillette and reads her nudity as a mirror image of the chaotic portrait. Instead of the fullness of the multiple, she is the model for its blankness, she is the starkly possible.[12] "She is the bare faculty, undetermined, undifferentiated" (36). Having called chaotic fullness by the name of "la belle noiseuse," Serres now names blank and undetermined multiplicity a "dereliction." Forgotten and weeping, but loving, loving Poussin to the point of reluctantly agreeing to exhibit her beautiful nudity, Gillette is "[t]he dereliction of nudity, the dereliction of the possible or the dereliction of the origin, the dereliction of freedom" (36). Serres's repetitive use of "déréliction" carries that much more weight for our reading of his exposition, as it refers in English, in addition to a state of abandonment, to "a recession of water leaving permanently dry land" (*Webster's New Collegiate Dictionary*).[13] *La belle noiseuse* as "the beauty of the waters" is now repeated by the inversely complementary blank "beauty" of dereliction, Gillette as the *absence* of the waters.

Serres's reading of the double aspect Balzac offers of "noise," noise and quarrel on the one hand, silence and blankness on the other, underscores the all and nothing, the undifferentiated and illogical "middle" of possibles that Serres attempts to understand pre-phenomenologically. These are not contradictory states, but two expressions by which we may understand the multiple, all *and* nothing, not all or nothing. This double hearing of "noise" is pursued further in two chapters, "The Ballet of Alba" which we may call a "chapter on Gillette" or on "silence/blankness," and "Collective Furor" which describes the fullness of "noise," though each displays as an undercurrent the contents of the other. In the first of these two chapters (*Genesis*, 27–50), Gillette as blankness and derelict possibility undergoes a series of metaphoric transformations, each borrowing from literary, artistic, or historical sources, each adding to the process of *becoming* blank and naked capacity for the multiple. She becomes the joker who holds both zero value and all assigned values, absence of meaning and all meaning (40). Finally, her utter dereliction evokes a historical place that has no history, Alba, Rome's sister-city (44). (In Latin, white, pale, or blank is *albus.*) Forgotten and buried under the success of Rome, and under all the normative forms Serres attributes to the concept "Rome," Alba is the site of our undifferentiated, multiple origins, the "matrix" of our history.

In the companion chapter on "Collective Furor," Serres hears the multiple once again as "noise," but the image changes and smudges the reader's first impressions of a new notion of temporality. What Serres understands Balzac to be hearing in the dynamics of Frenhofer's visual creation, he himself hears now in the clamor and furor of history. The noisy twin and mirror of derelict Gillette/Alba is Rome, Serres's cantankerous paradigm of human relations. These passages are an echo of the Girardian argument that sustains *Rome: The Book of Foundations* and describes human social relations formed of rivalrous clamor and jealous quarrels, and the cyclical, eternal return of violence and death as a historical founding mechanism of Western culture. However, the process of "historical noise," and what it means in terms of human destiny as a moral question, is now modeled differently. In Balzac's story, the delicate, nude foot discovered in a corner of Frenhofer's destroyed painting is a first form, a first harnessing of multiples into an ordered construct, something both painters understand and welcome as a harbinger of the "known." Their attention to this phenomenon in the midst of *la belle noiseuse* is the very reason that Gillette as the model of utterly blank and chaotic multiplicities is forgotten. Balzac's story visibly separates the multiple into two models, as cantankerous "noise" and derelict silence.

In Victor Hugo's poem "Les Djinns," Serres discovers the same process in the model of a "silent" and undifferentiated whisper that swells into demonic noise and clamor, engulfing the poem world with deadly violence. But, in the same way that the nude foot can be extracted from the portrait of chaos and become knowledge only when it becomes the repeated and ordered choreography of steps in a "ballet of Alba," Serres hears the chaotic waves of infernal din and demonic noise in "Les Djinns" break before the wall (*le mur*, in French) of ordered existence. There, the poetic I begins to harness the whisper into repetitive little noises (*le murmure*, in French), which can then be manipulated as historical noise and grow into a crescendo of murmurs, rivalrous noise. The nude foot, which becomes the dancer's repetitive steps in a quest for artistic purity of form, is analogous to the violently swollen Hugolian whisper, which breaks against the wall of human resistance and becomes a repetitive echo of information. In both cases the multiple possibilities of the chaotic Ur-noise are harnessed into a static order to arrive at a concept of time channeled into the portrait of a historical foundation.

The argument of historical noise, however, reconnects time to a phenomenological and anthropological context repeating the themes of *The Parasite* and *Rome: The Book of Foundations*. Here, where the goal is to arrive at time freed of such contexts, it risks losing its pre-phenomenological origin. The contrast between the uninventive but ordered repetition of the Same and the open possibilities of the chaotic multiple rests entirely on a play of words (*mur*, *murmure*, *écho*) which takes advantage of a linguistic coincidence in the French language. The *mur* of the Hugolian poem as an ordered construct of human intelligence offers protection from the onslaught of the "Ur-noise" which remains frightening and incomprehensible, like Frenhofer's chaotic, unknown masterpiece. The écho of Ur-noise bouncing off the *mur* creates *mur-murs* which will eventually grow into the clamor of history.[14]

Hugo's poem, of course, does not overtly present such a historicist interpretation; it is a poetic mood piece in Hugo's collection *Les Orientales* which reflects the contemporary period's heightened interest in orientalism. Through his oblique reading Serres adds a dimension that creates a model, not of history's beginnings, but rather of the underlying causes of historical time understood as inherent strife, violence, and recurrent mayhem. Based on a thin layer of lexical or quasi-anagrammatic resemblance, Serres's logic walks a fine line between what Hayles terms "wacky theorizing and brilliant insight" (207). One could argue, however, that the poem itself is wacky and that the mythical notion of a "djinn" itself is a vague and unsubstantiated fantasy. A djinn, Hugo' poem, and Serres's tenuous juxtaposition of wall (*mur*) and repeated whisper (*murmur*) are examples of models by which reality and our experiences with/in it are simulated. As such, Serres's strange twists and logical bends are no less valid than a whisper coming out of the desert and growing into the murderous din of a million evil spirits. They are radical in the sense that they attempt to wean us from a unitary, separate understanding of the world of djinns, or Hugo's poem, or the painted form in the corner of a senseless portrait, and to steer us to the recognition of an identical Ur-"noise" that foments in the foundations of all our discourses.

Serres's unique and odd perspective on Hugo's "Les Djinns" is valid only as a poetic model that simulates an evolving process. His objective, however, is thereby rendered much clearer: our common concept of time as history is "noise" safely covered, is disorder masterfully harnessed. Rational thought is but a putting in place (thesis), and then—in direct contrast to Frenhofer's mad method—a repetition in place (the static) until it mushrooms into the strife-torn din of history and the triumph of science.[15] Both domains obey and support rationalism's redundancy until it reaches its Hiroshima and brings us to global violence. But "Les Djinns" suggests another outcome, namely, that the crescendo of noise can also diminish after having reached its absolute fullness and slip back to a whisper slowly evaporating into the vast space or "noise" whence it has emerged. Hugo's poem lacks a demented Frenhofer who would recognize and celebrate the creative "vapor" of the irrational (74).

Hugo remains a powerful influence on *Genesis*. From "Les Djinns" Serres learns that turbulent Aphrodite does not repeat her steps in place. Neither is she portrayed as chaotic madness. She foregoes her place and leaves repetitive clamor behind (74–80). She leaves behind those who "put into place" (thesis) and others who dispute the right to the same place (antithesis). She is the excluded middle who is defined by a "foregoing" and who, by her ceding movement, begins a *process* (77). Serres has thus traced the notion of time without recourse to concepts that imply a historical dialectics. But he has to admit to a double time. One is static, a repetition of action; the other is dynamical, always ceding and moving to an unknown space. One is the historical time of human relations with its inherently violent cycles of renewal that strive ineluctably toward a unitary goal at the expense of possible variations. The other is time as flow or ceding-the-place, the turning and walking away from the place (*cedere* means to go away in Latin), a meandering fluid time that Serres will later rediscover in his reading of Ulysses's travels.

The two ways of seeing time, static and dynamical, become the subject of "La naissance du temps," the most profound and haunting chapter of *Genesis*. First, Serres gives a Girardian outline of history as mimetic rivalry, which is the demonstration of what I characterize as "static" time (81–95). As before, his model is Venus, but now la "belle noiseuse" reveals her noisy, cantankerous nature and becomes the "hideous noisemaker." The second part of this chapter deals with the process of foregoing cantankerous rivalry. Since it means "leaving the place" of static redundancy, I shall call it "dynamical time." Serres now suggests that we "think time, think process, directly" (97). Modeled in Balzac's story as chaos, time has no determined direction of flow, while time as echo (Hugo's poem) has a direction that we can count on and are already attuned to, because it is repetition. In order to understand (hear) chaotic time, Serres must "explain" chaos as something other than disorder. Heard/understood outside the realm of concepts, "positive chaos" (98) is now presented as an open system, a multiplicity, a fundamental or "basic time" (100). It consists of non-calculable, non-integrable fluctuations each of which is an unexpected jolt or incentive for the next; but none ever repeats the exact characteristics of any other fluctuation in this "chain" (71) of time. From time as imitative echo with a unifying future, we go to a time as fluctuations with unpredictable possibilities.

The reflections on both repetitive, directed echo and non-integrable fluctuations prepare a notion of time as invention, time of the tiers-instruit in a later text; yet they seem unable to resolve the dilemma of a double understanding of time in *Genesis*. Serres therefore shifts his discussion to the sciences (103). But the author of *Hermès*, whose work is marked by a global plan for closing the artificial gap between the humanities and the natural sciences, faults the sciences he loves so much for forgetting their foundation as a process of transformation (101). Instead, they increasingly imitate the furor and jealousies of human relations (104). Time as a nonlinear dynamical process is forgotten in favor of historical time, even by the sciences. Instead, Serres proposes inventive reasoning as a scientific mode of thinking which proceeds with scientific rigor as well as liberating refusal to obey the safety of imitation (105). The scientific mind must rethink a turbulent state, intermittent between the unitary and the multiple, between the solid and the fluid (106–110). It must think non-repetitive time as a mapping instead of a linearity (115). This is a reference to topology, often called "rubber sheet geometry" by scientists, as the best way to understand time: *La belle noiseuse* is such a topological portrait of time. Echo as an epistemological concept has no place in this portrait, except perhaps as a secondary sign, like the foot poised for its first step, in order to put the process of knowledge into place. When we focus on the foot alone, like Poussin and Porbus, trying to reconstruct the full portrait from one detail, we cut short all possibilities in favor of one possible step which then mushrooms into its own repetitive echo. It grows into sameness, in linear fashion: the principle of identity is midwife to linear time and knowledge (119). Serres wishes to reconcile the understanding of the function of the foot with the chaotic multiple of the rest of Frenhofer's canvas. This reconciliation is the fine line of a phase transition where turbulence combines elements of order and disorder. The highly inventive areas of transition are referred to as circumstances (120), an expression that wonderfully combines the regularity of the -stance or static with the topology of the possibles, the circum-. This lexical image will become a leitmotif in the last pages of *Les cinq sens*.

Genesis is the discovery of a dilemma. The multiple is the object of this text, history its objective (7). Obviously, the two are meant to have time as a common nexus. Yet Serres arrives at a double hearing of time, time as pure multiplicity, and history as multiple possibles redundantly reduced to unitary order. In contrast, the Hugolian reading of chaos, wedded to Balzac's story, implies time as a *pure* multiple "ceding the place" to the furor and noise of rivalries and violence. For Serres, historic time that is founded on redundancy and mimetic repetition is above all *stasis* or static invariance, to which, in the past, he had often assigned the statue as a perfect conceptual metaphor reinforced by etymological kinship.

Now, he repeats this image: History is a sum of stabilities, "stupid, heavy . . . statues, [like] sandbags on the ground" (120). He prefers to search for what he ambiguously calls a "new stability, a secret one, perhaps unheard of," and which he hopes to find in transformations produced irregularly by circum-stances (120).

Two observations can now be made concerning Serres's double vision of time. First, both the static and the "dynamical" are expressed by "stability," the latter by a "new stability." Second, the statue as the paradigm of the stagnant static that Serres consistently aligns with death as its telos, is contrasted to a turbulent state that is "a median state between a slightly redundant order and pure chaos" (121). Complex dynamics theory calls this "deterministic chaos." Hayles calls it more dramatically "chaos bound," like an echo of "Prometheus bound." For Serres, this state evokes a turbulent Venus, the Anadyomene surging from the waters but still covered by them, still of the "noise" (*la noiseuse*) but already forming herself (*la belle*), either to step (or run) in place (static) or to cede the place and pass on to chaotic circumstances (dynamical). Turbulent Venus "is the first form, [she] is the first tower" (122). In other words, she is a turbulent statue.

The double perspective on time can therefore be stated in the following way: Time is considered under two aspects, the static and the dynamical, historical time and the "fundamental time" of the multiple. The enigma of time is thus not really resolved. History as a destructive time is described by "noise" deformed into rumor and dull repetition. On the other hand, "noise" of *la belle noiseuse* is turbulent Venus as a first form or statue (tower), a turbulent statue promising dissipative possibilities. *Hermès IV. La Distribution* expresses the same idea of a double form, but still in the scientific language of information theory: "The senseless din is made meaningful by [a] series of rectifiers" ("The Origin of Language"; Harari and Bell, 81). In *Genesis*, the static as stagnant redundancy and the dynamical aspect of turbulence are hesitatingly wedded in a statue which turbulently rises and fades from the reader's grasp: a fluctuation, an oscillating state between the pre-phenomenological and the phenomenon. Armed with the chaotic power of *la belle noiseuse*, the reader will not simply see Venus, but turbulent Venus—a model which gives us a glimpse of time as such.

The surprise of *Genesis* (and its built-in weakness) is this glimpse of a phase juncture at which the static and the dynamical coincide. What is only an ambiguously metaphorized notion in this text, will later become the discovery of time as a dynamical force at the very heart of the static itself, in *Statues*. *Genesis* does not patently arrive at that crucial insight, although it is there latently. Written from a Girardian perspective, this text does not take advantage of the dynamical process discovered in the notion of *la belle noiseuse* where the argument comes tantalizingly close to revealing a "new time." Instead, Serres falls back on the issue of "noise" as rivalrous furor and repeats the knowledge of time as a historical concept of cyclical exclusions that act as guarantors of the cultural phenomenon.

In the last chapter, "Rêve," Serres returns to the image of turbulent Venus as a first construct. The tower of Babel is here the paradigm for an implication that stands the biblical metaphor on its head: "Babel is not a failure; it is at that very moment when the tower is dismantled that we begin to understand that one must understand without concepts" (123). The Bible does not mention the collapse of the tower, but attributes the failure to complete the project to the confusion of the builders's language (Gen. 11:7). The people cannot communicate any longer about a construct that was meant "to make a name for [them], so that [they] may not be scattered about the whole earth" (Gen. 11:4). Their desire for order and control envisions a tower that would connect their lives with the heavens, individually as well as globally. This is a perfectly reasonable undertaking, equal to artistic representations of feminine beauty epitomized in Venus, equal to the expectations of two painters in fervent anticipation of a privileged viewing of a masterpiece. The Bible tells how God foresees that this unique undertaking of "a single people with a single language" will lead to absolute dominance rivaling divine power itself (Gen. 11:6). "There will be nothing too hard for them to do," God says. This linear, power hungry development is arrested by divine fiat. Since then Babel has forever become a paradigm of confusion and an example how not to structure society. But for Serres the unintelligible chatter at the foot of crumbling walls is intrinsic incompleteness, it is a "non-integrable multiplicity." Chaos is not a failure, "incompletion is the ordinary state of affairs" (*Genesis*, 125).

Not the towering peak, but the inchoate babble of human relations is the sod out of which we begin to build with concepts, "a set of distinctive objects, pieces, sections of Babel" (127), little Babels that will rise and fade in a chaotic djinn-ballet of the "Discernible" interspersed with the "Indiscernible." For Serres inverted Babel is the Leibnizian pyramid inverted (127–28). No longer is the unifying peak the ideal rising from the fog of disorder, as in the *Theodicy*, for it is merely perfection, it is complacently at peace. Serres has understood Balzac and Hugo: only the "multiple," imperfection in the sense of non-completion, and the confusion of words do not cut short the richness of possibles. Only in this richness lies hope for inventive thought (132). The problem of evil is here already implied as a question of linear exclusionary thinking.

The god of the Bible was right; he stood at the city gate and knew what had to be done: the word Babel does not only mean confusion but also "the gate of the god" (Gen. 11:9n).[16] Serres's inverted Babel is the gate that permits us to espy chaos and order all in one, ordered reasoning with open-minded invention, a true masterpiece harboring human capacities that, if *heard*, will move us as close to the divine as we can ever hope to be.

2

Time Immortal: Reading *Détachement*

> Only in Marco Polo's accounts was Kublai Khan able to discern, through the walls and towers destined to crumble, the tracery of a pattern so subtle it could escape the termites' gnawing.
>
> —Italo Calvino, *Invisible Cities*

In the final chapter of *Genesis* which he calls "Dream," Serres describes an epistēmē founded on "noise" as chaos and music. *Detachment* (English translation) is a kind of decoding of this dream through other dreamlike tales, with chaos doing its own story telling. Subtitled "Apologue," this slim volume avoids conceptual argumentation and demonstration in its implied objective of linking the solution to the problem of evil, a human historical problem, to a new way of understanding and living time. It buries history (in the flood plains of China, part 1) and frees the human subject from jealous relations. Detached, he can fly and soar across the landscape of both space (part 2) and time (part 3). It is a didactic fable that finds—the stuff of fairy tales is made of this!—ancient signs of immortality.

Rome: the Book of Foundations and *Detachment* were both published in 1983. From the chronological distance of our reading, it appears as if the two notions of time, one stable and deadly linear, the other unstable and chaotic, were translated into two separate discourses. *Rome* is a history book, a historian's nightmare perhaps, a history book nonetheless, in the fashion of the boustrophedon writing of the ancient Greeks Serres had learned to read in the company of Hermes and the parasite.[1] It offers a functional description of time conceived historically, and arrives at an epistemological description of societal foundations. The "other" time needs detachment from stable forms in order to come into "view" (discussed by Serres in *Les cinq sens* as a dynamical "visiting"). *Detachment* is a personal history, without being autobiographical. Nature and art, not human relations with their often cantankerous and hateful din, provide the topological spaces where the narrative subject finds a past that is a stage setting for an immortal future. Immortal, time sheds its last distinctions: past, present, and future become one. The "oneness" of time is discussed much later by Serres and Bruno Latour in *Conversations on Science, Culture, and Time*.

Rome is a case study of the Girardian thesis of triangular desire as it is formulated in *Deceit, Desire, and the Novel* and globalized in *Violence and the Sacred*. *Detachment* detaches itself from the deterministic aspects of this thesis. In *Things Hidden Since the Foundation of the World*, Girard posits a resolution to the endless cycles of sacrificial violence. From the perspective of his theory, the messianic message focalized in the historical figure of Christ calls for an imitation to end all imitations, that is, to renew the "old Adam" in view of a historical redirection void of any sacrificial or exclusionary compulsion. According to Girard, Christianity has never truly committed itself throughout its long history to such a reading of the "Imitatio Christi" as the only valid message of salvation. In *Detachment*, Serres describes a different solution to the murderous noise of human relations. He calls for a renewed experience with nature, with the objects of the physical world as they are in themselves. Primary focus of the later study *The Natural Contract*, this appeal aims here at an empirical approach in order to loosen the ties that bind his epistemology to a Girardian reading, as we have seen in some parts of *Genesis*. "Detachment" as a title applies therefore to its methodology as well as its content; it constitutes an epistemological review of the relationship between the human subject and his world. One can argue that part 4 of this apologue, with its story of Diogenes, again describes in a generic mode what *Rome* does for a precisely determined historical locus and era. It, too, reveals the victimage mechanism as a chain of repetitive violent exclusions, thereby rendering its beneficial agency ineffectual. But in the process, it relinquishes all ties with time identified as a process of human work and returns to an encounter with the world of objects. We shall see how Serres deconstructs the role of the Girardian victim by the strategy of a dynamical reading of human interaction with *natural time*. This allows him to propose a *prescriptive* process, albeit in his own radical way of understanding the legal notion of prescription, as a genuinely peaceful substitute for the sacrificial crisis mechanism crucial to Girard's philosophical anthropology. Alluded to in *Detachment*, prescription later plays a paradigmatic role in *Le Tiers-Instruit*.

With *Detachment* Serres's discourse bifurcates into possibilities of reading: it is the fading echo of a historical reading; it is a synaesthesis that leads to *Les cinq sens*; or it is a fundamentally different, dynamical approach that will come to full recognition in *Statues*. The focus of its empirical reading is the "space" of the world that eventually translates into a renewed contemplation of history. It prepares the way for *Statues* which discourses on this new mode of freeing time from its historical straightjacket so that its recorded expression in the human sphere of action may be redefined along lines that are parallel to Prigogine's "new time" of science.

The first part of *Detachment*, called "Farmer," describes a visit to the floodplain of China stretching from the delta of the Hoang-he and Yangtze Rivers to the loess-covered fertile hills. This vast plain has been covered with the handiwork of the Chinese farmer: the visitor experiences the historical takeover of the earth by a totalizing agriculture that leaves no speck of soil unused or misused for frivolous purposes. This "absolute" agricultural policy has nothing to do with the given political system but denotes a historical takeover of millennial duration. Repeatedly, Serres describes fields that blend into each other without any demarcations. There are no border strips or soil banks, where nature could remain undisturbed, in a marginal mode, the way it does at the edges of tilled acres in the French countryside. They are silent markers of nature as yet untamed, the unruly "rest" that can be cultivated—perhaps—at a future date. But in China, agriculture is an inundation: "I am not speaking of the act of flooding of rice paddies, but of cultivating the land. . . . [Chinese] farming has covered over everything, like a tidal wave. It is the totality. It is so positive, so rational, so well-adapted that one can only speak of it in negative terms" (5). These words set the tone for the rest of this part 1. For, what Serres writes of the cultivation of the soil could also, indeed should also be read, at one and the same time, as the "culturization of the Earth."

The tableau Serres paints of the countryside covered by humankind's cultivation/culturization is not meant to be a lesson on Chinese life. Juxtaposing space and time, or nature and human history, while his eye canvases the featureless floodplain, Serres experiences what it means to have reached the end of historical time. The rivalrous cycle of foundations on which *pax romana* is constructed here gives way to the exhaustion of history. Like their houses from which no paths seem to radiate in welcome movement, Chinese farmers are mired in the vast expanse of the cultivated loess. Here, time no longer flies, it slows down; time stands still. The end of history is a static time, spatially demonstrated in the perfected geometry of the fertile plain. Reason has optimally landscaped the earth until it has become stagnant, monotone, death-like in its absolute stability and totality. Here, Serres sees the attainment of all terms, equilibrium at its most perfect state, reality experienced exclusively as rationality.

And Serres experiences anxiety, discursively visible in *Detachment*. The passionate geometer in him agonizes over the sterility produced by geometric perfection in its linear rush toward unitary mastery of space. In the *Hermès* series, especially in *Le passage du nord-ouest*, he had repeatedly written about the origin of geometry as one of the most fertile strata underlying all of Western epistēmē. It is interesting in this context that ten years after *Detachment*, Serres publishes a study the title of which shifts from the singular "origin" to the plural: *Les Origines de la géometrie* (1993). What the apologue in the guise of fables, parables, and stories tells about the creative vivacity of multiples and fuzzy mixtures, the 1993 text presents in an epistemological survey, in which Serres comes to terms with the curse inherent in a singular, linear view of geometry, "the perfect realization (*la réalisation parfaite*) of our mastery over the world" (*Les Origines*, 133). But if we recognize multiple origins, we also understand that "this new knowledge breaks ranks with modern philosophy which evolved from medieval scholasticism and Cartesianism and was founded on the opposition of subject and object. [The new science], on the contrary, is comfortable with the pronouncements of Anaximander [the pre-Socratic philosopher of the *apeiron*, the undefinable or unlimited]" (*Les Origines*, 105). It is therefore not surprising to find echoes of *Detachment* in *Les Origines de la géometrie* (60–64) where Serres paraphrases from his own apologue to underscore that "as of now the old order of things has been upended (*Nous venons de renverser ce vieil ordre des choses*)" (*Les Origines*, 133).

Detachment, however, describes Serres's anxiety as a parable. The difference in style leads to an apparently different story, but the reader should not be detracted by what in this reader's estimation is one of Serres's most poetic texts. For as such it is also profoundly scientific in the sense of science as knowledge of the world. Serres experiences anxiety. Son of peasants, he begins to tremble at the thought that his ancestors' art is inexorably condemned onto the same rational course. But then he jubilates at the thought of all the brush-covered strips separating French fields, the grassy banks along the paths cutting across the cultivated valleys, these "useless" remainders of nature still left untouched, where weeds and brushes grow and bloom in disorderly, haphazard, "uncultivated" fashion, the residue or lack of rational work, the imperfection that still holds forth all possibilities. He recognizes the lack of perfection and the discrepancy or deviation (*l'écart*) from absolutely rational culturization as the *rest* that keeps his world—the farming of his ancestors and his own activity of writing—from becoming ossified, from becoming too perfectly stabilized on the vast plain of human labor, be it fertile land or a written work. The as yet uncultivated field borders or the half-finished abandoned page give hope for possibilities and future inventions. "This swerve brings time into existence, it produces it. Something remains for an upcoming project" (9). Serres hopes that the lack or non-perfected residue will not disappear, for it guarantees a time as yet "unused."

"Of what can we write a history when all the goals have been reached, when a perfect balance has been secured . . .?" (10). Culled from the monotonous horizons of China, Serres's question is the concern of contemporary science. Prigogine and Stengers, for example, propose the examination and evaluation of far-from-equilibrium systems as optimum systems with which we can adequately model the complexity of our world. Their "dissipative structures" allow them to interact with their surroundings in dynamical, constantly changing processes which produce time. The authors "emphasize the unexpected character of this behavior" and point to nonlinearity, instability, and fluctuations as key traits (Prigogine and Stengers, 12–13). Serres's anxiety is driven by the knowledge that the eradication of imperfection by rational labor means loss of movement, or possibilities cut short, to fit a measured, rational, evolutionary straight path. "A touch of irrationality is a saving grace for us, a stroke of luck which gives us some breathing space, a loose fit in the machine which makes us alive. . . . Perhaps we shall learn one day that the most reliable machines leave room for the unexpected" (*l'imprévu*) [22]. The reflections on the cultivated countryside, both

Chinese and French, echo a profound epistemological understanding of the essential value of imperfection as the guarantor against deadly stagnation befalling human inventiveness. True invention can only thrive on a dynamics of the unpredictable. Imperfection means far-from-equilibrium conditions which, in turn, mean movement, both entropical toward equilibrium and dissipative toward self-organizing new structures. But movement, in turn, implies the notion of time.

The fable of the Chinese farmer, however, has a sequel which tells of a spatial configuration that may accomplish a new perspective on time. Time is taken out of its historical framework and aligned with the workings of nature, but without any implication of a reassignment to the temporal reversibility of classical scientific models. This constitutes a crucial difference. In order to understand what Serres is getting at, we must backtrack a bit. In *Mathematics and the Unexpected*, the mathematician Ivar Ekeland outlines the fundamental changes the Keplerian revolution forced upon the way scientists modeled the world. His principal description is consecrated to the "figures of time" that scientists have since then attempted to trace out into comprehensible shapes. This slender volume presents a critical survey of the spectacular advances made in scientific descriptions of time. In the course of his own efforts "to portray time" which is an "evasive model [that] leaves only a facet of its rich personality at each sitting," the author stresses again and again the role of the "unforeseeable" (*l'Imprévu* of the original title in French) in the dynamical models that modern science relies on to scientifically calculate the complexity of nature (Ekeland, 120). The paradoxical meeting and interaction of calculation or measurement (the quantitative) with unpredictability (recuperated in qualitative approaches) are the hallmark of complex dynamics that underpin Ekeland's "figures of time." Serres searches in vain for similar markings of time, as his writing eye surveys the Chinese floodplain. "Meaning in its wholeness covers the land without any waste. What is missing is nonsense" (*le non-sens*) [*Detachment*, 11]. It is clear for the author that the Chinese farmer is not the only one buried in this rational loess, for our technological and economically driven world inexorably demands the same slavish and exhausting fidelity to rationality.

In *De l'imperfection* (1987), A. -J. Greimas, otherwise well known for his structural semiotics (*Sémantique structurale* 1966; *Du sens* 1970), identifies three "axiologies" on which our system of values is based, the epistemic, the ethic, and the aesthetic. He assigns to the aesthetic axiology the workings of imperfection, because only the aesthetic value resists categorization and subordination: "[Aesthetic] judgments are not balanced (*équilibrés*), they do not divide values into positives and negatives" (85). The aesthetic experience always means a transgression of the rational frontier (23–33) and a confrontation with or escape towards "the Immanence of the perceived" (72). "At stake is our paltry everyday life; at stake are the various means by which to introduce fractures into it" (95). These fractures are the signals of imperfection; they allow us to experience "aesthetic seizure" (*la saisie esthétique*) and that "illuminating fascination (*l'éblouissement*) when things fall from perfection into imperfection" (13).

Greimas's argument parallels Serres's, but the metaphor of falling is inverted in *Detachment*. The result, however, is the same: Serres refuses to succumb to the unstoppable trend toward rational saturation. After all, he claims, when he tills the soil, only his feet are planted in the ground, and perhaps his hands through his tools, but the rest of his body stretches upward, vertically toward the sky.[2] There he discovers many signs of time, because changing weather phenomena occur in all directions (time and weather, *le temps*). Dwelling in the sky, his body is no longer chained to the earth, he rides with the winds, he flies, he lives by the wind and the weather, like a mariner at sea. He experiences the fascination Greimas found in aesthetic rapture: For the latter, the enchanted fall into imperfection is the leap to a new meaning, analogous to Serres's "new knowledge." "Imperfection appears to be like a trampoline projecting us from insignificance towards meaning" (le sens) [*De l'imperfection*, 99]. But this meaning transcends what rationally makes sense. It is also more radical than Serres's initial change of direction to the vertical which still leaves his feet firmly planted in, and anchored to the ground.

But Serres returns to the Chinese farmers and discovers in their customs a crucial difference. They have invented a wonderful—ecstatic—way out of the quagmire of their ordered timeless lives. It vaults them into an existence of "perfect" imperfection. In their leisure time they, too, surge upwards into the heights of the sky, for they are the children of the dragon, ancient symbol of China. As such the dragon is most often displayed as flying dragon-kites. Without a doubt, Serres is referring to Kites' Day, the Teng Kao or Feast of High Flights, when all of China flies aloft, in the form of millions of flying dragon and other favorite animal kites dancing in the air. There the people find movement; there they are no longer prisoners of a frozen history. Up among the winds, the clouds and the unpredictable flux of air currents, imperfection is regained, for the flight of the dragons cannot be predicted absolutely. Even the most skilful hand at the lines does not know when and how to anticipate the next gust of wind. For Serres the spectacle of a billion dragon kites dancing in the wind and floating towards the sun is emblematic for a people that is soaring away from its exhausted land (16).

The story of the levitation of a whole country, its people and its architecture (the roofs of pagodas curling upwards), is a Serresean parable that helps us perceive topology in a nonmathematical fashion. In this description, the images of pagoda roofs and kite lines emphasizing the purely vertical give way to, and blend into a floating "plane" of fluid elements, wispy clouds, curly smoke rising from chimneys, gusty breezes, and the dancing tails of the kites covered with writing in the Chinese tradition. It is on this airy plane, composed of fluid images of turbulence, that the Chinese peasant escapes the cultivated perfection of the loess-covered plain. The vertical movement alone will not bring liberation from the geometric grid covering our culturized plain, since our feet are deeply rooted in it; it merely adds a third dimension to what remains essentially a Cartesian structure. Instead, the Chinese leave the geometric plain with its linear vertical extension, to live on a topological plane of fluid shapes and unstable conditions. They have acquired a cultural wisdom that compensates for their agricultural perfection (38).

Serres's fable describes, in fact, what is meant when topology is called "rubber sheet geometry," and topological transformations are likened to deformations that a well-defined area on a "sheet" undergoes when the latter is stretched and twisted. Topology has become an increasingly important tool for dynamicists because of the possibilities it offers to visualize the unpredictable behavior of nonlinear dynamical systems, through multidimensional spatial simulation in virtual space. When the Chinese fly their dragon-kites, writes Serres, their dreams fly with them, they rise and fly in that fluid plain/plane. There is nothing miraculous about this transformation, nor, for that matter, anything imaginary about dragons. Western rational thought does not apply to Chinese wisdom, which allows them to dance on the fluid plain of clouds, while a million feet are rooted in the fertile loess of the floodplain. From robots obeying the rules of ordered rationality, they emerge as acrobats free to sway and dance on the ever shifting clouds in ecstacy. Their steps are never *exactly* repetitive; they live a dynamical system. Flying with their kites as sails, the Chinese are mariners of the sky who have revived their history and restored time dynamically, without noise.

Part 2 of *Detachment* is dedicated to the "Mariner" and transposes the flight of the Chinese to Serres's own shores and into the lives of people who left indelible marks on his past. It is the story of a friend who once was a sailor, like Serres. He detached himself from land, set sail, and rode the winds and the choppy seas, both topological plains where dimensions—and thus perspectives—are in a never ceasing flux. Once the mariner has lived through a perilous storm, his life can never be the same, for he has learnt that the most solid and invariant experience of life, the weight of things, becomes an utter variable in heaving waves and on storm-tossed crests of water. In heavy seas he may feel near-zero gravity or crushing g-forces. He understands the variable character of his world. Perched in the crow's nest high up on the masthead, he perceives space itself to be changing topologically, as length, width, and depth are forever varying, stretching, disappearing, and reappearing (73).

Both the mariner and the farmer fly. In their wisdom, they know how to detach themselves and encounter ecstacy in the imperfection of the variable. Serres translates the wise imperfection he perceives in the high-flying Chinese dragon "dance" into a kind of sainthood for his French mariner. The fabulist's logic is impeccable: The dragon-kites are the Chinese's intermediaries between the ordered, coordinated world of the fertile floodplain and the variable elements of the sky. They teach the robot farmer how to soar into liberating imperfection. On the other hand, angels are traditional intermediaries between the heavens and Serres's world (ancestors, friends, himself), but they play the same role as the flying dragons. When the mariner floats and dances in the topological space of the sky, he flies and dances in the dwelling place of saintly angels: "Yes, holiness is to fly" (34). For Serres, detachment is imperfection is inventiveness is sainthood. It is the manifestation of a dynamical understanding of space—with a promise of new insights into the mystery of time.

Part 3 is perhaps the most significant of the four chapters comprising *Detachment*, because its story is a series of discursive flashbacks with temporal zigzags that project time as immortality and thereby already allude to the juxtaposition of time and death as the dynamical force of *Statues*. Its title "Wanderer" is only one of several meanings we should read into the original "Errant," a seeker, an adventurer, a lost soul, one who goes astray, or one who misunderstands. Each announces an aspect with which we can unlock the implications of this text.

Trees served as catalysts for the upward motions in parts 1 and 2; now once again, a tree becomes the symbol of a verticality that reaches for a liberated state posited by Serres as a necessary condition for invention, the true basis of knowledge. It was at the moment when the fabulist's eyes perceived a mulberry tree at the horizon of the floodplain, adding the vertical dimension necessary for the upward stretch, that he recognized the Chinese dragon-flight as an escape from "an adapted eternity" (*une éternité adaptée*) of reasoned conformism (10). Similarly, Serres saw his sailor-friend climb one of the tall stone-pines the region of the "Landes" in southwestern France is famous for, in order to prune dead branches and safeguard his house and family. At that moment, he recognized that his old friend was still flying in and with the winds, as he had done when he had climbed aloft into the riggings and took in the sails to save his ship from going under in stormy weather. For Serres, this dynamical experience of life means that, actually, he was always flying in the company of angels, like a saint.

In part 3, these trees of particular insights become more generally trees of knowledge and trees of life. The biblical book of *Genesis* tells of the tree of knowledge in the Garden of Eden and of its fruit which, once eaten, would give knowledge of what is good and what is evil, a wisdom God intended to reserve for himself. But the Bible also speaks of a second forbidden tree, the tree of life equally off limits for the created human, because its fruit, once plucked and tasted, would give immortality. God exiled Adam and Eve from Eden so that, having once disobeyed his law in order to gain divine status, they would not attain immortality as well (Gen. 3:22). How Serres *recuperates this immortality* without invoking God's wrath, that is, without transgressing into the divine, but by *inventing it within the finite human sphere*, is the aspect which makes the "Wanderer" section of *Detachment* so epistemologically profound. It lays the groundwork for a new concept of time.

First the fabulist relates dreaming of a tree that was a chaotic mixture of plant and animal forms. It was at once monstrous and richly alive. In this way, Serres recreates a tree of life of a purely natural but chaotic form, an animal-plant on which, he speculates, he could graft himself if he were to reach out and let his arm become one of the branches (43–44). The Edenic tree of life acquires here the concrete, material aspects of animate life—plant, animal, and human animal. But it is a dream and the vegetative life thus envisioned does not convey any potentiality for life without death, that is, potentiality for immortality.

Therefore, Serres wanders on and exchanges the dream for a mythical scene set in Roman antiquity, in the mountains of Alba. There, the ancients worshiped the goddess Diana in a sacred forest on the shores of Lake Nemi. One tree, in particular, was said to have magic powers and was fiercely guarded by a priest-warrior against anyone who tried to approach and pluck one of its magic boughs. The Golden Bough and its legendary magic bestows upon its owner all power over death and grants him the secret of life. What in the dream was a mere possibility of a peaceful union with the vegetative tree of life, becomes now a question of conquest: Who will possess the secret of life and all its concomitant powers? By whatever title myth and history acknowledge the guardian of the sacred tree, priest, warrior, king (or scientist), "he spends his life fighting a deadly war. Thus the war unto death is the sought-after secret. What foolishness!" (46). Serres reads the Roman myth as a version of our cultural origins from which our traditions have descended under the sign of rivalries and envy. Our cultural din overshadows the real world and its objects; the secret of life falls into utter forgetfulness. The Edenic tree of life has changed into a tree of rivalries under the branches of which we play out a life of combat for combat's sake: "The secret of the golden bough is being lost . . . for never having been looked at. Thus, the secret of knowledge is destroyed in the know-how of the quarrel. . . . We turn our back to knowledge in order to fight face to face" (48).

The golden bough of life has become a tree of mortal combat. The over-rationalized and perfect plain of Chinese (agri)culture finds its analog in the tree of death. Its dry branches stand for the ossified knowledge of inimical rivalry; it no longer signifies liberating verticality. We can now state the moral of Serres's story as it has unfolded so far: He has learnt that life and knowledge are fruits of one and the same tree. The secret of life is knowledge of time, and the knowledge of good and evil is the wisdom to live in a time that is not a historical sequence of cantankerous "noise," but time as fluidity and imperfection, as possibility for invention, for peace and tolerance.

And Serres continues his quest for a fluid, inventive knowledge. Exchanging dream and myth for artistic expression, the wanderer comes upon a painting by the fifteenth-century master Petrus Christus, the *Madonna on a dry tree* (49). The painting depicts the Virgin Mary with the Christ child in her arms, standing in the fork of a thorny withered tree. Bathed in light reinforced by the vivid colors of the woman's blood-red cloak and the whiteness of the child's clothes and complexion, the figures stand in stark contrast to the dry, bare tree and the black background. Here, the tree of life offers nothing but branches ending in rigid spikes and long thorny outcrops. The dreamer's animal-plant tree of life, transformed into the golden bough of Roman mythology holding the secret of life long since forgotten, now is a dead thornbush offering spikes of petrified, hardened ideas. It has become a tree of death. The tree of knowledge is presented as a withered tree of monotonous, repetitive ideas and cliches of lifeless and uninventive knowledge ensnared in competitive and rivalrous battles for power. The thorny spikes are visible symbols of the weapons of this war.

Of course, Serres is not per chance grafting an epistemological meaning onto a late medieval depiction of a withered tree from which, seemingly, springs forth a new branch or blossom. He is aware of the rich symbolism embedded in the portrayal of the Rosa Mystica, one of many epithets bestowed on Mary; he understands it as a representation of the "root of Jesse" and the "branch of David," messianic symbols that are frequently used in the prophetic books of the Old Testament. The Messiah is the promised branch of life to which the Middle Ages, dedicated to a flourishing Mariology, added Mary as the mediatrix who makes the reblooming of the branch possible.

But Serres's eye is not so much attracted by the light emanating from the Rosa Mystica in the middle of the dead thornbush, as by another twinkling fluid light that comes from fifteen golden letters *a* in gothic style, which gently sway like leaves suspended from the dead branches. They refract their golden light at varying angles and thereby appear to make the branches vibrate: the withered tree is brought back to life with its tiny golden bells and becomes a Golden Bough once more (95). Serres believes that the painter combined Christian and pagan ideas to depict the renewal of life. But he does not exploit further the symbolism of the painting and leaves unexplained why the letter *a* was chosen to represent the burgeoning life of the tree, and why there are fifteen such letters.

I would like to suggest a further interpretation of the symbolism in order to demonstrate how extraordinarily felicitous Serres's choice of this particular painting is for his argument. In *The Waning of the Middle Ages*, Johan Huizinga describes the eccentricities of fifteenth-century religious "symbolism in decline," which was the inevitable consequence of a whole era in decline. Among many examples, Huizinga details medieval prescriptions for the use of the rosary with its repetitive prayer offered to Mary. The number fifteen, for instance, carried great significance, because the Ave Maria, the first part of the (now) Catholic prayer known as Hail Mary, consists of fifteen words in its Latin version (Luke 1:28.42). They represent the fifteen virtues or perfections of the Virgin Mary, each of which drives away a sin when emulated by the faithful (Huizinga, 208).[3] In the painting, the fifteen letters *a* distributed in the branches around Mary and the child clearly indicate the fifteen words of the Ave Maria and echo the famous first word Ave of the archangel Gabriel's greeting. For the fifteenth-century artist they also symbolize the fifteen mysteries of the rosary, which celebrate the fifteen ways to attain eternal life.[4]

The dead tree of the painting is thus transformed before the eye of the viewer into a symbolic carrier of life, with the fifteen Ave's representing its rejuvenated branches or the steps of a staircase that leads the virtuous believer to heaven and eternal life (Huizinga, 208). The tree of death is transfigured before our eyes into a tree of life or a staircase leading to sainthood. We have closed a gap in Serres's text: The vibrant Marian tree of blessed life is the pine tree in the crown of which the old sailor flies with the winds into sainthood; it is the dragon-kite in whose company the Chinese farmer flies into the ecstacy of inventive imperfection.

For Serres, however, the three fables—the dream, the myth, and an artistic expression of religious symbolism—do not suffice to tell the story of a new time. He needs one more piece to complete the puzzle, that is, to discover the new temporality as an ancient wisdom culled from forgetfulness. He turns to nature and enters the giant sequoia woods of Northern California. Singling out one tree for its "colossal, monstrous" majesty, Serres stands in awe-inspired terror before this "orant" which has been stretching to the heavens for over four thousand years, continuously. In contrast, our culture is focused on death and forms a concept of *death as a function of time* in an entropic equation that thermodynamics formalized into a science. Our culture is built on an erroneous philosophy of negatives and our rituals on bloody sacrifices. The sacred awe invading the quester comes from the realization that he stands before a living being of nature, of brute but true nature such as we do not know her anymore, since our rational totalizing work has cultivated her. This being lives through the ages; it does not know death. Time expressed as history and destruction here gives way to a time revealed in nature, in the sanctity of a giant old tree (53–55).

The tree is so extraordinary that Serres imagines it did not grow there by accident, but was planted, found, or invented; or perhaps it was formed, modeled, pruned, and cultivated by human beings many thousand years ago. Hypothetical and Rousseauistic as this Genesis-story may be, it tries to express the extraordinary quality of *life as a function of time* that our prehistoric ancestors must have recognized and nurtured into sacred rituals around the giant sequoia.[5] Central to this fable is the idea of a quest for a long-lived species that could perpetuate human prayers to the gods. The ancients found it in a tree representing life expressed as *time going toward immortality*. By recognizing its indestructible, millennial, sacred nature, they invented time as a giver of life. They planted a visible, living promise of their own immortality; the giant sequoia is a message to us from an era that had the wisdom to cultivate life and invent time. But our rationally dominated culture has "brought death into this garden, into this banquet of immortality" (58). Serres's awe before the colossal life-tree derives from the insight that the ancients knew how to live immortality.

The fable of the Californian sequoia is not a story of angelic travel through a topological space, as in parts 1 and 2, but more specifically of flight through time. Perhaps the most fantastic of the apologues in *Detachment*, it is nonetheless the more convincing in the "topological sense" that Serres employs here: the depth of his awe persuades him that any ancestor must have felt the same inspiring reverence in the presence of this guardian of longevity. In a sort of flight across time, he fictionalizes a prehistoric ritual: the celebration of immortal, "living" life that takes the mortal sting out of human finite existence. Flying back in time gives him an understanding for taking flight into the future as well, where he hopes to arrive at a new understanding of history evolving toward living immortality.

Perhaps this fable seems too fantastic for our rationally conditioned imagination. At most, our static perception admires the stupendous beauty of the sequoia and—ever the cataloguing and quantifying mind—counts its growth rings as proof of its unbelievable age, but it is incapable of soaring topologically back and forth through time. Therefore Serres gives other examples of longevity taken from cultural activities: the domestication of animals and plants, the invention of writing, the development of logic, arithmetic, and geometry. Serres rightly wonders why we call them useful inventions, which is a purely static perspective; we should think of them dynamically as signs of our immortality, since, once invented, they never disappeared but flourished into multiple cultural expressions. We allow ourselves only to consider their utility and forget that the genius of prehistoric ancestors has expressed and preserved in their enduring longevity the gift of our immortality (60–61).

The absence of dynamical thinking is a latent theme of Serres's apologues which try to pry us loose from our rational bog, detach us from our self-produced bonds, and put the wind back into our cultural sails, our kites. Our perspectives are fundamentally marked by static thinking which Serres sees in today's drive for instantaneous results and gratifications. An idea born today is dead tomorrow; our culture thrives only under the aegis of death. But this is not the final statement of part 3. In one of the mental twists so characteristic of his writings, Serres adds an ironic correction: "We also invented immortality. Neither in written signs, nor for the living, immortality on the contrary is in lifeless things. Nuclear waste that we do not know where to put will surely last more than four thousand years" (61–62). The bitter legacy of our rationalized, technologically enslaved generation to some misbegotten descendant is a deformed immortality, foreshadowed by the stockpiling of radioactive waste, as deadly, stagnant, and horrible as the "end of history" that stomps the Chinese into the immobility of his absolutely perfected landscape. The moral of the fable is that we have to become dynamical; detachment is the only means by which we can reverse the trend toward stasis and so perhaps save our descendants from our own nightmarish perversion of immortality.

The fourth and final part of *Detachment* is also its longest. Called "Franciscan," it seems to shift its focus, away from both the predominant images employed so far and the epistemological pursuit of a new perspective on time. Its subtitle expresses three characteristic objects of human relations, "Wagers, Fetishes, and Objects of Exchange"; they form the fateful attachment or tethering by which Serres pessimistically describes cultural stasis and the disappearance of inventiveness. Although the exposition of the argument resembles the Girardian anthropology of cultural foundations, its strategy is different. In his inimitable fashion, Serres retells the story of the Greek philosopher and cynic Diogenes who goes to great lengths to avoid any relations, disown all objects of exchange, reject all fetishes, and decline any wager. He is the saint who leaves the illusory world of competitive relations for the real world of objects, animals, and love. The story of Diogenes demonstrates how one can overcome the curse of mimetic desire that Girard identified as a fundamental trait of the human psyche.

Serres's version of the Diogenes story concentrates on the cynic's desire to find and experience the pure objects of empirical reality stripped of all interfering tools made by men and subject to their coveting desires. Naked Diogenes and stark reality form "the crystal of dereliction" (67), when he experiences water, snow, and sunlight, as detached but real objects. He becomes Gillette's twin and shares her dereliction so that he can meet "reality as such" without the stultifying interference of concepts. Unlike Poussin and Porbus who fail to recognize beauty offering itself (herself) in its pure, unfettered, "derelict" nudity and look in vain for it in a representation, Diogenes rushes to embrace, for instance, the derelict solidity of snow. Here, Serres's meaning of detachment is honed to its barest expression in the union of subject and object.

Among the many Serresean statues, paradigms of stability but also of sterility, Diogenes' snowman acquires a special significance. Serres describes it as a statue (68), but does not comment on its possible stasis because his attention concentrates on the gesture of derelict embrace. He does not dwell on the resulting unity of subject and object, for his intention is to portray the cynic's famous quest for a man as a quest for a "lost object." It is a search for the real world to which belong possible human relationships without war (wagers), competition (objects of exchange), and adoration (fetishes) (69).

Something very akin to the dilemma of a double perspective on time occurs at this moment in *Detachment*. In *Genesis*, a book on time by Serres's own description, he opts for a Girardian reading that forces him to conclude that we must contend with two visions of time, ordered time of history and a chaotic dynamcial time of invention. Unable to reconcile the two, *Genesis* ends on a note of promise. Serres's failure to fully recognize the wisdom encapsuled in the creative madness of Balzac's old fool is mirrored in *Detachment* where he passes over the scene of the composite statue of Diogenes and the snow sculpture. When Diogenes embraces the snowman, some snow certainly melts under the warmth of his body and covers both the sculpted figure and the man. The icy droplets cover both and make them one, one fluid system. The embrace produces fluidity; once again a Venus Anadyomene is in the process of rising from the waters but still engulfed by them. Wedded to the snowman, Diogenes is "Vénus turbulente" implied as equivalent to a chaotic state that promises unlimited possibilities for creative invention. This encounter sets in motion a fluid process of "multiples" that is ignored here but will be addressed in *Les cinq sens* and reformulated as a dynamical process in *Statues* and *Le Tiers-Instruit*.

Serres tells another story, however. Diogenes is experiencing the world by the simple act of absorbing sunlight, when Alexander the Great passes by and parasitically plants himself between the philosopher and the sun. Quite simply, Alexander functions as the biggest parasite to impose his laws on the relationship between subject and object. Serres favors the meeting of Diogenes and Alexander, because it represents the absolute example of two mutually exclusive experiences. On one side is the man who is closest to successfully avoiding all interfering manipulations, achieving nearly zero level of attachment and a minimum of need, allegorized in Diogenes. On the other side of this scale rises the densest screen, the deepest shadow between subject and object, the most glorious domination by a different sun, the maximum of force arrived at by war (wagers or risks taken), by the bartering of values (the usurpation of the market place), and by prosternation demanded (the status of a god-like fetish), all personified in the great Alexander (70–71). In the parasitic hierarchy of relations that humans erroneously, and catastrophically, substitute for real objects, "royal" power of the absolute magnitude of an Alexander is the "object" most idolized, fought over, and bargained for. This paradigmatic parasite produces a maximum shadow, precisely placed between the real object, metonymically represented by the sun, and the human subject detached from all rivalries. It prevents all possibilities for a "new knowledge" arising from the union of subject and object, and for a "culture [that] constructs the set of means which permits escape from comparison . . . which does not delight in death, [and] which laughs at hierarchy" (72).

Diogenes, the maximally disinterested one, is pitted against the maximum interest, Alexander. The full impact of this encounter is felt, when Serres reminds the reader that "interest" means precisely that which resides between, that which places itself squarely in the interval (132). Diogenes, on the other hand and because of his disinterest, achieves the perfect union of object and subject without interference. "He is the inventor of the fundamental theorem of knowledge. . . . Diogenes says: things to see, things to know, water, snow and sun are uninteresting" (73). Unencumbered by interests which invite rivalry and competition, Diogenes is free and detached to know genuinely. Serres promises his reader that this fundamental disinterested knowledge, unscreened by and unfiltered through desires for power, position or economic considerations, will make inventions pour out onto our world. Presented in a fable, this is the theorem of Diogenes. It makes possible the "instructed-middle" of *Le Tiers-Instruit* whose trademark is an inventive and liberated spirit, and where the para-sitos passes to the "instructed-middle."

But the almighty Alexander holds a trump card in the calculation of maxima. As the absolute parasite who conquers all, he cannot tolerate that even the least of his subjects escape the force field of interests. Such a refusal or evasion would unravel the absoluteness of his status. Therefore, the absolute possessor offers Diogenes anything he desires. We all know the cynic's answer; Diogenes requests of Alexander that he step aside: I ask nothing but that you get out of my sunshine. This exchange is the point of departure for a Serresean train of thought that is emblematic of his brand of baker's logic and that will color the remaining pages of *Detachment*.[6] The new epistemology, Serres writes, is "a hymn to the sun. It is, basically and beforehand, a matter of displacing the king's body" (78). But since all available space is always already the king's space, Diogenes' request amounts to an act of opposition to and revolt against the king, against his maximum order and his maximum power. Opposing the maximum parasite is falling into the latter's trap: the philosopher himself enters the struggle for power, by a dialectical set-up that cannot be circumvented.

Serres is quite specific in explaining that he presents the case of Alexander as a paradigm for a general condition or law of absoluteness we all are subject to. Only the paradigm changes now and then. Serres does not hesitate to equate the absoluteness of the ancient king with the modern tyranny of the atomic bomb, the absolute modern power which has invaded and filled all space: it is our nuclear Sun-King. Here, this title takes on a derived meaning, more powerful than the historical Sun King ever possessed at the pinnacle of his reign. Absolute power at a maximum, whether king or the modern divinity called Science, recuperates even the sun. Even if he (or it) wanted to, he could not step aside, because he is everywhere, omnipresent, and omnipotent; the king's shadow falls everywhere, absolutely. Resembling the paradox of trying to oppose Hegelian dialectics and by this very opposition falling into its unavoidable, logical web, Diogenes' request is a "catastrophe" in view of his quest for a detached existence. In fact, Serres speculates that he may have planned this meeting in order to provoke the king into a display of his absolute power—the better to display his own weakness, powerlessness, and dereliction (81). His role as "victim" of absolute power is so perfectly executed that Alexander supposedly expressed the desire to be like Diogenes, if he were not Alexander. If he were not king, he would like to be a philosopher; and if he did not hold power, he would like to be a cynic.

In the meeting between Diogenes and Alexander, a relationship of need and desire is uncovered that mirrors the mechanism of mimetic desire with its attending force field of rivalries. Diogenes is the Girardian victim who, by his self-imposed public exclusion, guarantees the social order expressed by the royal power. In a Girardian reading of this scene, Serres declares the cynic the winner in the sense of a victim who is elevated to an indispensable position of care and respect that may reach divine status. But he is equally a loser, because he succumbs to the siren call of rivalry: he competes with the Sun-King for a little sun.

This is a startling development toward the end of a text that so beautifully fables or "fabricates" possibilities for detachment. But Serres's strategy is simple, although intricate: "[Diogenes'] slovenliness becomes a weapon and his dereliction a decor" (82). In other words, Diogenes is a miserable Alexander whose protestations of disinterest and dereliction are as suspect as proclamations of methodical doubt by staunchly Cartesian traditionalists. For he remains ensnared in the thicket of relations; "he clings to maximum stakes, he manipulates solar fetishes, he creates value" (82). Alexander and Diogenes are forever bound as two rival poles in the triangular scheme of mimetic desire, of which the third is the desired object of absoluteness, be it in the guise of power or dereliction. To show how pervasive this Girardian schema really is, Serres cites three other examples which are not readily associated with mimetic desire and victim-victimizer bondage: The tragic conflict between Antigone and Creon, La Fontaine's fable of the lamb and the wolf, and the story of Job confronting God.

In the case of Antigone and Creon, love is pitted against hate. For over two thousand years, Antigone the victim receives our pity and her power of love our admiration, but Creon is delivered up to our hate. The victim triumphs over her victimizer; and yet, she has lost, because love turns into an assassin, over and over, on stage. "This love which kills every evening must be another name for hatred" (83). Serres sees no real difference between Diogenes and Alexander, or between Antigone and Creon. The victim becomes the "victimizer" for having succumbed to the temptation of vengeful competition. The same pattern holds for La Fontaine's fable. The wolf deserves our disdain for so flagrantly flaunting his power and demanding his right. But every time this story is told or read and our pity showered upon the innocent victimized lamb, the wolf turns a bit more irrevocably into a scapegoat, the big bad wolf whom everybody loves to hate. Serres ends his demonstration with the story of Job, the victim on a grand scale, who claims the right to scold God, his victimizer, and thereby usurps divine judgment. The common element in these examples is the exchange of roles, or rather the equalization of the Girardian victim and victimizer. Emblematic for all victims, Diogenes plays the game even better than does Alexander. Or as Serres writes: Against all odds, "Diogenes knew how to move the body of (the) Sun-King" (88).

In the Serresean version of Girard's "sacrificial crisis" or "crisis of indifferentiation," variations become immobilized and impossible, and the force field of relations settles into an ordered space of power. In the last pages of *Detachment*, we are no longer reading apologues illustrating the exhilarating freedom of far-from-equilibrium states. Serres laments their absence, when he speaks of relationships of desire that loop back onto themselves and stagnate into relationships of order. "The relation of order produces values, unleashes wars, it is the great fetish, the matrix of fetishes" (92). Order is the fetish we all strive for, because we consider it a moral necessity for which we even go to war. It is our flood plain over which we bend in endless labor, mired in the servitude of mimetic relations.

What Serres describes here is the reality of our attachments. After all, whenever he speaks of detachment in the first three parts, he fables and fabricates an ideal human condition that our reality is as yet awaiting, and for which he hopes to facilitate the (re)discovery. And yet, even within these last passages lies buried the hint of a "new" time. The notion of "prescription" is a deconstructive gesture by which Serres tries to detach his discourse from all historical strings with their Girardian perspectives. However, before considering the notion of prescription and as a preparation for a full understanding of what is at stake, it is useful to recall one more story of a victim-victimizer bond, not told by Serres. Baudelaire's poem "Abel et Caïn" (*Les Fleurs du mal*) fits perfectly into the series of exemplary stories briefly described above. It relates a classic case of mimetic desire, using a setting as old as the biblical Genesis, but also projecting elements of Serres's argument. The poem progressively depicts how the victim Abel turns, through his descendants, into the paradigm of the bully, while inversely the murderer and victimizer Cain grows into a victim whose descendants deserve pity and empathy instead of our condemnation.

Every time the Bible story is read, Abel points an ever harsher accusatory finger at his assassin; every time Cain becomes that much more guilty, his crime a bit more heinous. Analogous to the roles that Serres's perspective assigns to Diogenes, Antigone, the lamb, and Job, Abel enters into a rivalrous competition with Cain. The innocent victim condemns and assassinates the latter with ever greater success, as the word of the Good Book spreads. What Serres understands as happening on stage every night the Sophocles tragedy is performed, or what is awakening in the reader's mind when he reads La Fontaine, or what is coalescing into moral virtues when Diogenes is admiringly quoted, Baudelaire sees happening every time the sacred text is opened for spiritual reading. Therefore, he retells the story of Abel and Cain in such a way as to avoid adding to the *static* repetition of putting Abel on a pedestal, like an icon. Instead, he tells the story of their descendants; he addresses alternately the "race of Abel" and the "race of Cain" and records what happens to them *as time passes*. This dynamical approach gives a different slant to the Biblical story and precludes a mere repetitive "staging" of the innocent victim and his victimizer. On the contrary, Cain is exonerated through the working of time understood as a dynamics of imperfection: "Race of Cain, your work is not sufficiently done."

Baudelaire, of course, has his own message of revolt: Abel and his descendants create their history by engaging in fetish worship, enriching themselves through shrewd exchange of valued merchandise and usury, and warmongering. Abel's fat, historical "body" will reach its end of time, not as a martyred saint but as a cadaver good enough for making the earth where it will be rotting belch a stench-offering to the heavens. But Cain's time is not composed of competitive relations; for him and his descendants time comes as a thousand uncertainties that never allow them to sink into monotone stability. This time does not permit their work or suffering to come to a finish, it resembles time as pure multiplicity, which Serres discovered in Balzac's story of Frenhofer's unending refinements done to his masterwork. Far from the historical repetitiveness of Abel's race (which "pollutes" the earth with its greedy swarms, says the poem), dynamical time gives credence and value to Cain's act, in the eyes of Baudelaire.[7]

It is the absence of dynamical time that Serres laments when he contemplates the historical and mythological forces which shape our epistēmē. What Baudelaire expressed in his poem appears quite innocuously as "prescription" in the last pages of *Detachment* and is easily passed over for the author's urgent appeal to his reader to see Diogenes, Antigone, La Fontaine's lamb, and even Job for what they are: victims turned avengers many times more ruthless than their original victimizers. Prescription is primarily a legal term which can be variously interpreted as the effects of statutes of limitations, of squatters' rights, or of the forgiving of debts. But whenever and however prescription establishes legal rights, it is always based on time, a time which has the power to undo or alter previously established laws, that is, the power to undo the work of historical time. Since it plays a significant role in the development of Serres's argument in *Le Tiers-Instruit*, I cannot end my reading of *Detachment* without pointing to its importance in allowing Serres to dream again, at least in the very last pages, of unlocking our lockstep culture.

Lamenting our epistēmē's attachment to a thousand ties of its own making, Serres imagines lifting anchors and flying or sailing away with the Chinese farmer, the sailor friend, and the ancients of the sequoia forests. Dynamical time is the secret of their ancient wisdom. We, too, could make it our own, Serres suggests, if we were to give prescription a chance to play out its power. Then, mighty Alexander would not look quite so ridiculous when facing the helpless but defiant Diogenes. But ignoring prescription makes Diogenes into a second Alexander, more ruthless and mightier in his cynicism, because we stubbornly repeat his sarcasm for two thousand years, while Alexander cannot help being himself. The same goes for Creon who was driven by political necessity to a hateful act. But Antigone becomes a worse Creon, a thousand times over, when we permit her to condemn him to our hate and spite, at every performance, all in the name of love (87–88). Implied, of course, is the possibility that love would and could not turn to ruthless, millennial hate, if prescription were allowed to do its work and Creon's act seen as one hateful moment within the flow of time. Then, Creon could be forgiven and Antigone's love freed of repetitive "noise." If prescription were not so utterly ignored, those who tell La Fontaine's fable to their children and grandchildren would understand that the wolf follows his natural instinct and acts like a bully. The lamb turns out to be much more dangerous because it acts like a victim yet victimizes the bully by delivering him to our hate, over centuries, and within the web of human relations depicted in the fables. "The wolf is the real invisible victim of the real wicked lamb. The lamb is the real wolf" (85). Serres is quite rigorous in his critical view of our repetitive ways of conceptualizing historical time. It fosters a kind of hatred that "is beyond any comparison with the simple and savage hatred of the wolf who satifies his hunger" (85).

Serres regrets our culture's neglect of prescription and our forgetfulness of its epistemological power. By ignoring it, we tether ourselves to the closed order of rational social constructs (167). We make of our history a static time of uninventive repetitions of the Same, with prescription relegated to the juridical domain. Instead, we should invest in its dynamical properties that could help us exit the circus of our tightly dovetailed attachments (172–73) and the theater of our never-ending rivalries (175). Prescription would enable us to leave, set sail, fly, go astray, and experience the objects of this world. That possibility becomes the subject of *Les cinq sens*.

3

Time Empirical: Reading *Les cinq sens*

> A 'black box' is a conventional agreement among scientists to stop trying to explain things at a certain point.
>
> —G. Bateson, *Steps to an Ecology of Mind*

As presented in *Detachment*, Serres's epistemological critique implies the reappropriation of a dynamical understanding of time. The detachment necessary for this fundamental process is fabled as a soaring from the furrows of our fields of life towards a topological space of multiples. There, unattachedly meditating on the objects of our world and how we fit among them—relate to them instead of reducing them to calculated investments in our cantankerous relationships with one another—constitutes a first ahistorical step into time. This double gesture, retreat from noisy competition and experiencing the immediacy of the world, is replayed in the opening scenes of *Les cinq sens*. This work is but the first step in the evolution of Serresean discourse towards an epistemology of dynamical time that will come to full expression only in *Statues* and *Le Tiers-Instruit*. The all important first step, however, is the subject of this chapter.

In one of the best synoptic reviews written on *Les cinq sens*, Thomas Kavanagh emphasizes that its five chapters "do not, in any simple way, correspond to a cataloguing of the five senses" (Kavanagh, 938). Rather, they present a demonstration "that our sense of ourselves as 'subjects,' in spite of Socrates' endless words and Descartes' careful *cogito*, is grounded in and cosubstantial with our physical bodies as the seats and receptors of our five senses" (Kavanagh, 938). Serres maintains that we have forgotten this fundamental relationship of one-ness, that our sensate bodies have become separated from our being as subjects, that the data of the physical world, filtering through our senses, no longer reach our intellect in their full immediacy. A screen has been put between the two, which we call "language." The screen prevents passage *and* it allows passage; it bends, distorts, refracts, or alters what passes; it is a filter of words and signs through which we become aware of and experience the objects of the physical world—but screened.

Les cinq sens is therefore above all a text on the complicated yet simple relationship between the senses and language. It reflects Serres's grave reservations about a long Platonic tradition in philosophical systems and his implied criticism of its modern expression in France via phenomenology and its mutations.

Semiotics, in particular, with its categorical disavowal of referent and subject in favor of an independent textuality is implicated whenever Serres addresses the problem of language. Yet, there are moments in *Les cinq sens* when he seems to make an exception for his own text and assign the power to create reality to his own words. These instances not only give them their rhetorical vigor and power of persuasion, but they also leave the reader bewildered when she is urged to reject the alluring "drug" of language and empirically experience the world, by a text which claims an exception from this advice for itself. Bruno Latour discusses a similar dilemma for scientific literature when semiotics is practiced on scientific texts which by definition must be referential and should be transparent as to their language. However, Latour points out advantages that can be gained from a semiotic analysis; then, the scientific text "becomes an event, which has the same activity, the same materiality, the same complexity, the same historicity as any other event" (Latour 1992, 130).

Analogously, Serres's text with its appeal for an empiricist approach to reality may be considered such an event. Its aim is a new adventure in philosophy that goes against all traditional thinking since Plato and seeks to reestablish a place for the sensate experience of the objects of our world. For, writes Serres:

> All around us, language replaces [sensate] experience. The sign, so soft, substitutes itself for the thing, which is hard. I cannot think of this substitution as an equivalence. It is more of an abuse and a violence. . . . My book *Les cinq sens* cries out at the empire of signs (*Les cinq sens crient à l'empire des signes*). (*Conversations on Science, Culture, and Time*, 193).

This project seems doomed to failure, especially when we consider recent advances in computer technology and developments in virtual reality. Indeed, Serres arrives at a point where he feels the need to defend the very language, whose despotism he decries, against new developments in the information sciences that seek to encode the world of objects in a pure knowledge based on algorithms, models in virtual space, and digital data banks. The screen of words and signs becomes now itself blocked by a new and more powerful screen of abstractness that will demand and necessitate computer literacy but leave us doubly forgetful of the physical world. For Kavanagh, Serres's

> attempt to write on the five senses takes on all the tragic poignancy of a quest for an object which is doubly lost. . . .
>
> Left with a language which science has stripped of all pretensions to represent the real, Serres embraces it as a medium now cleansed of all functions save the esthetic: that of celebrating the beauty of the world perceived by our five senses. (Kavanagh, 940–41)

This convoluted bond of complex relations between subjective intellect, language, and the physical reality of objects is frequently modeled in *Les cinq sens* as a black box with input and output, exactly like scientists do when they "stop trying to explain things at a certain point" (Bateson, 38). Whenever Serres speaks as a scientist here, he posits the physical world as input for the black box, the latter as unknown space of sensate variability or sensation, and the information flowing out of the box as output which we as subjects work into a meaningful epistēmē (136). Serres the scientist goes no further.

However, Serres the poet-fabricator goes beyond and attempts to enter the black box where, of course, explanation is not forthcoming—or it would no longer be a black box—but where he experiments with "implication," to arrive at a different understanding, a step on his way to a "nouveau savoir." To grasp the precise but multiple meaning of "implying," I want to linger at one particular passage of the first chapter in which the author reflects on the sense of touch relegated to the shadows cast by the privileged position of the sense of sight in Western philosophical tradition. Among diverse examples that serve to develop a "philosophy of the sensate," he discusses the tapestries of "The Lady with the Unicorn," an ensemble of six individual pieces dating from the late Middle Ages, which are today the central attraction of the Museum of Cluny in Paris. Serres goes beyond a pure description of the tactile and essays to lay out programmatically the process by which we render abstract what originally faces us or opposes (ob-ject) us in its physical concreteness (52–60). In other words, his strategy is similar to that outlined at the beginning of *Genesis*: We must uncover the pre-phenomenological movement in which object, subject, and language meet and blend before they separate into their phenomenologically respective domains. Once they are split, we neatly categorize them as input and output unexplainably linked by a region we conventionally call "a black box," thereby handsomely and somehow logically exhibiting our embarrassed ignorance.

Serres follows traditional wisdom when he interprets the themes depicted on the tapestries as illustrations of the five senses and the sixth tapestry as an allegory of the internal sense. He points to certain spatial differences in the tableau of the sixth tapestry depicting a tent pavilion, which evoke this internal domain of the sixth sense (53). At the same time, it summarizes the other five by bringing them to a closure through a movement towards internalization in the represented scene (55). Here, the woman is the "resultant" of her other portraits. Standing in the doorway formed by the pulled-back drapery of the tent, she appears to be both inside the tent (her upper body) and outside on the flower-strewn grass (her lower body and the train of her flowing robe), an impression reinforced by techniques of perspective typical of medieval visual art forms.

The tapestry's double aspect of the outside/inside is only obliquely hinted at by Serres, but it is useful to point out the visual perspectives of the sixth tableau before attempting to follow Serres on a difficult journey into the black box. He begins by calling attention to four words inscribed on the ornamental edge of the tent's canopy: A MON SEUL DESIR. Since the rediscovery of the tapestries in the 1840s, it has generally been assumed that a certain Jean de Chabannes commissioned them as a gift for his fiancee, the beautiful Claude Le Viste, member of an influential patrician family in Lyon. According to one version of the tapestries' vague origins, the inscription, the only words represented on the six tableaus, expresses admiration and passion for his betrothed.

Serres, however, reads this inscription with the Latin origin of the word *désir* in mind, and so understands it not only as an expression of longing for a beloved, but also as a term which has retained the meaning of regret. The tent, and the woman (half)-entering into it, depict above all the body turning inward and closing upon itself while regretting what it is leaving behind. During this process the lady touches, smells, hears, tastes, and sees, at the same time that the words on the canopy betray her nostalgic longing for what has already been forsaken (57). We may understand this scene as analogous to *Vénus anadyomène*, with the full implication of the adjectival attribute discussed in chapter 1.

The jewels depicted on this tapestry symbolize the objects of sensation represented on the other tapestries. The woman is about to return them to the open jewelry chest held up by her lady-in-waiting. Or so Serres interprets this gesture which occurs at the very threshold of tent and grassy island, that is, at the demarcation line between inside and outside, between the inner and outer *sensoria*. But this betweenness is also the space where language (the four words) appears:

> This is the very first sentence, the initial proposition, as original as the original sin committed by a woman on an island-paradise, as original and permanent as hers. Written [on the canopy] are the first words which the body acquires when it turns inward upon itself, speaks, cloaks itself with tongues of fire, and produces signs. . . . Here is the sentence that makes the world of objects retreat. . . . Here are the words that drive the senses back into the black box. We desire nothing but the reopening of this chest. (57)

From this tapestry Serres implies the triple coincidence of (1) the emergence of desire as a fundamental, nostalgic remembrance of an absence; (2) the perdition of the world of objects to the immediate grasp of the subjective intellect; and (3) the birth of language as an expression, revelation, confirmation, and relational medium for the first two. The only way to experience, even remotely, the objects of the world is now a path through language. The original written proposition encircles the tent-body which closes upon itself and forgets the outside (57).

Most observers understand the tableau of the sixth tapestry differently: Stepping out of her pavilion, the woman is about to remove from the chest a richly bejeweled collier, presumedly a wedding gift from her betrothed. None less than the poet Rainer Maria Rilke explains the scene in this way, in *The Notebooks of Malte Laurids Brigge* written during the years he spent in Paris as secretary to the sculptor Auguste Rodin, at the turn of the century. His interpretation of the tapestries is purely impressionistic and descriptive, but his introductory remarks allude to something lost of which only a nostalgic longing remains. Without explicitly hinting at the sensate, Rilke foreshadows in poetic prose what Serres implies in *Les cinq sens*. He describes the tableaus as woven "pictures . . . which give praise to everything and give nothing up" [*Bilder . . . , die alles preisen und nichts preisgeben*] (Rilke, 131).[1] He continues with a parenthetical remark on the fate of poets who created only with language and then falsely believed they were doing so in a more literal and precise fashion (*wörtlicher*). In reality, poets are never allowed to know more than language can give (Rilke, 131). As Rilke's spiritual godchild, Serres laments the appearance of language as simultaneous with the disappearance of the immediacy of objective reality.

But there is more to the spiritual and poetic bond between Rilke's reflections and the "philosophy of the sensate." Rilke observes young women, touched by the first stirrings of our century's emancipation movement, who seem to be drawn to these tapestries like to a magnet. While contemplating the woven beauty, the women seem to "forget themselves a little," and the poet realizes that they "have always felt that this existed—a quiet life like this, of slow, never quite clarified gestures, and they dimly remember that for a time they even thought *this life would be their own*" [emphasis added] (Rilke, 131). But then he regretfully adds that, after succumbing for a moment, "the young women suddenly take out a sketchbook and begin to draw" (132). What appeared to Rilke as a momentary insight into something beyond or prior to representation, on the part of some young women contemplating *The Lady with the Unicorn*, is analogous to Serres's black box which holds the secret of our sensate power to accede to the immediate world around us. But Serres is as powerless to open it, as Rilke is to enjoin the young women not to repress "the unalterable life that in these woven pictures has opened in front of them, infinite and unsayable" (Rilke, 133).

For Rilke of course, the tapestries illustrate Love, the kind of immediate and ineffable love that he believes only women can truly bear. One might therefore hesitate to compare Serres's black box of the senses with Rilke's "black box," which holds the mystery of a love that goes beyond even the most powerful poetic expression. Yet, further reading into Rilke's notes on the Cluny tapestries almost forces us to make the analogy. When the poet speaks of endless efforts put forth by women's love in past ages, he cites some exemplary women. These "powerful examples of women in love" [*die gewaltigen Liebenden*] (Rilke, 134) labored mightily in such deep affinity with love that they could not contain the love they had become themselves and walked through life engulfed in radiance, "*as if they lived with saints*" [emphasis added] (Rilke, 135). Similarly, in *Detachment*, dynamical time is implied as the topological plain of variables in the sky, where Serres's sailor friend dances in the company of angels and saints.

Juxtaposing Rilke's description and Serres's reflections, we come upon the same ineffable space of "unexplained gestures" or topological "variables" where both, great women lovers walk in a radiant glow, and sailors and Chinese farmers dance with the winds. No matter what name one gives them, all live in the company of saints. Love and sensate experience, as conjured up by the poet and the epistemologist respectively, are two sides of the same coin: sainthood as liberation from rivalry and envious desire, a sort of saintly rapture necessary for a life of dynamical time and creative invention. In the final analysis, these are the same in the Serresean philosophy of implications. Our detour through Rilke's *Notebooks* is useful for understanding the forms of implication by which Serres urges us to relearn the use of our five senses. It reminds the Serresean reader not to skip over and ignore the passionate gesture, akin to love, which gives its full dimensions to the epistemologist's quest for the true role of the sensate in our lives. At the same time, it explains a mysterious undertone characteristic of *Les cinq sens*, openly expressed only in its last pages and reflected in the statement: "To be free to write for a woman, never again to write against [the objections of] a proof-reader (*correcteur*)" (368).[2] I shall return to this point.

The secret to reading *Les cinq sens* translates into a renewed "love" for the sensate which is the opposite of a mere "cataloguing of the five senses," as Kavanagh so succinctly states in his book discussion. The only difference between Rilke's feminine love and Serres's loving appreciation of all the senses lies in their respective metaphoric expressions. Both see saintly love as the unique key which will permit entry to the black box, unhampered by the screen of language, "because only wrong is done in the saying" [*weil mit dem Sagen nur unrecht geschieht*], as the poet puts it (Rilke, 127). For the latter, poets and their words have unwittingly put up a screen that does not permit love to give witness to its prodigious power of turning a human being inward upon her or his *immediate* self. For Serres, the sixth tapestry of Cluny depicts this language screen, but in such a way that its very depiction, when reiterated by linguistic means in both Rilke's and Serres's discourses, is also a breech through the screen, an opening of the black box. It is a paradoxical opening, because the seminal words of the canopy imply both the retreat of the senses from reality and an expression of desire to open and free the senses (57).

Rilke witnesses this event momentarily, in the transfixed gaze of some young women. Not wishing to "do wrong in the saying," he does not speculate further in the vain hope of clarifying the "gestures"; instead he translates the visual encounter into a poetic praise of the mysterious power of love. Serres, on the other hand, *reads* the origin of language. Here, language interferes "always already" and displays its power to swallow up the sensate and render its effects opaque. It screens off all immediate, objective reality from our intellect allegorized in the tent pavilion with its canvas decorated with tongues of fire, symbols of the pentecostal miracle of languages. But by the same gesture, language offers us the means to express the longing for that which is absent. The longing (*mon seul désir*) is a filtering movement that will not let go. The lady of the sixth tapestry is therefore very appropriately placed on the threshold of the tent pavilion, in that region where she turns inward and outward, simultaneously or variably. She is *Vénus anadyomène*. In Serresean logic, she holds the secret of the word: It is not only an opaque veil, like the tent canvas behind which she is retreating, but also a screen that permits passage, like an unveiling, when the lady's unicorn and lion lift the canvas to permit her free exit to the outside.

The fact that Serres reads the tapestry in this fashion means that he has deciphered the secret of the nostalgic origin of language and thereby opened the tent, emblem of the body as a sensate receptor. As soon as he has marked all the implications of the inward movement, Serres reverses therefore the represented meaning and states unambiguously that the lady, in fact, steps out of the tent and retrieves the precious gems, objects of the outside world, from the open jewelry chest. She walks joyously in the garden of immediate sensate experiences (59). Viewing this scene through the eyes and language of the epistemologist of the senses enables us to participate in this saintly joy, as sensate beings while remaining creatures of language. The merit of Serres's approach lies in his double reading of the lady's gesture on the sixth tapestry. Like Rilke, who refuses to *explain* the represented scene, he *implies* a black box where things can no longer be explained and sense perceptions are somehow screened by language. But concurrently, he implies the opening of the box so that language will release the sensate experience of objective reality in its full immediacy. Serres writes here in the double role of scientist and poet-fabricator.

I have singled out this passage of the first chapter, because it implies the intense entanglement of the senses with language. The reader must understand the *paradoxical nature of the word* which fatefully mediates between the sensate experience of objective reality and the subjective intellect, yet is also the tool to overcome the very same mediation when it has ossified into an opaque linguistic veil, rendered even more impenetrable by supremely poetic efforts, if we are to believe Rilke. Confident, Serres confides that he has found the key to this paradox, although well-meaning friends had said it was impossible. At first he agrees with them, because we all live in the "tabernacle of language," like the lady:

> We have always lived there with her; we never left. We have never seen, known, nor understood the tapestry of Cluny. . . . I live in the tent covered with tongues and crowned with an inscribed banner. Those who live in the tent with me demonstrate with scientific rigor that nobody can leave, that nobody has ever left. You will find, they say, *no language to say or write objects*, flowers or fruits, birds and hares, sounds or forms, tastes and aromas, *no language to write or say the world* [emphasis added] before language emerged. All you'll find is a tapestry in the Museum of Cluny.

> You'll find yourself stymied (*forclos*). They are right. I can neither write nor say the five tapestries, because whatever I say or write I say only about the sixth. The event of language has taken place; we cannot change that. (58)

But because of their exclusive dependence on "rigorous" demonstration, "they" do not understand that the lady's exit from the house of language is also our exit and that we feast together at the table of physical reality (59).

Serres has understood this, because he forsakes rigorous explication and sees the implication of the tapestry and what our language has made of it. On her walkabout through the garden of sensate delights, the lady is accompanied by the unicorn, a fabled animal and somewhat of a monster. She leaves the tent of language without forsaking the language which floats or dances, that is, those linguistic forms like fables, for instance, whose meanings are to be grasped topologically, not geometrically. Similar to the figures of the Chinese farmer, the sailor, and the seeker of immortality, of *Detachment*, who all accede to a topological space of variables, the lady is forever walking in the realm of the fabulous, of topologically variable language. For we know her only by a fabled name, as the *Lady with the Unicorn*. She is provisionally identified through a mythical beast whose nature is a chaotic composite of several animals; it is logically unclassifiable, and linguistically unnameable. Its "name" merely refers to an accidental, a horn as a singularity in the midst of chaotic multiples.

By way of implication Serres has "opened" the black box. Language can transfer sensate experience *directly*, screened yet immediate. In this process, it becomes variable and chaotic. The lady-of-the-fluid-language is the fluid composite of her five other portraits, she *is* therefore the sensate experience. But in the absence of precise identification and by aesthetic convention, we know her exclusively as the "lady with the unicorn," that is, we identify her with a construct that exists only linguistically (as fable or myth). The five senses are thus inextricably bound to floating, inventive language.

Serres's epistemological quest to reunite subject and object in an era of a new philosophy of temporality is thus confronted once more and decisively by the dilemma of language. Serres insists that we must face this hurdle, at the very moment in our epistemic evolution when language is threatened with enfeeblement and even disappearance. Kavanagh rightly identifies this insistence as an important undercurrent of *Les cinq sens*. It defies a growing tendency to relegate literature and the arts in general to the fringe areas of intellectual activity where they become "the noise of culture," as William Paulson very finely analyzed in his study by the same title. Against all odds, Serres puts his writing on the line and produces a language that speaks and writes (about) objects. With full knowledge that his friends are right when they categorically deny any possibility for success, he dares offer a linguistic guide to and through the senses. Since he has topologically seen and understood the tapestries of Cluny, he promises a beautiful gift of reality to his postmodern, information-rich reader.[3] To prepare this feast of joy, Serres "fabricates" the dense text of this book: "The senses, again at the start of yet another adventure, a phantom of reality timidly described in a phantom language, this I offer as my essay" (381). And in a final, humbly jubilant promise to himself as well as to his reader, he admits that, if successful, his efforts could lead to a glorious renewal.

The foregoing serves as an introduction to our reading of *Les cinq sens*. Once we understand the role of language in the (re)union of subject and object, we comprehend more specifically the validity of this text, perhaps the most densely written and most adamantly empirical of Serres's epistemological essays.[4] The actual introduction to *Les cinq sens*, and to its first chapter on the sense of touch, is a "painfully literal *in medias res*" (Kavanagh, 939), the tale of a sailor's escape from the hold of a ship on fire into the frigid, storm-tossed waters of the North Atlantic (13–17). The episode highlights a conceptual border zone that makes it difficult to distinguish where the subjective cogito ends and the objective physical body begins, and vice versa. Before considering the senses proper, Serres thereby posits the one premise on which the logic of his demonstration is based: The black box, in which the conversion process from sensate experience to intellectual comprehension takes place, is nothing but a fine line, and then often a fuzzy one at best. But it is everything, since the line is the screen of language and we are creatures of that very language. We are anesthetized by words and signs, as the reflections on *The Lady with the Unicorn* underscore. The linguistic drug makes us hallucinate that this fine line is a chasm so large that only language can bridge it. The double role of language as a pharmakon, both curse and cure, is Serres's initial overriding concern, based on the premise of the forgotten union of the cogito and res. The argument of its duplicity also serves, not so incidentally, as the ultimate defense for the linguistic construct that is *Les cinq sens*, because the latter could easily be labeled a classic example of an ironic undertaking.

In the context of the conceptual border zone allegorized in the story of the ill-starred mariner, I speak of a fuzzy line. The term "fuzzy" is deliberately chosen, not for an impressionist effect, nor for joining other critics who do not always agree with Serres's paradigms when these jump too boldly—or recklessly—over accepted borders that separate the domains of knowledge. I do, on the contrary, mean to underscore the rigorous logical and scientific discourse that takes place in *Les cinq sens*. Fuzzy logic and fuzzy theory are new conceptual tools that developed alongside nonlinear dynamical systems theories. Today, logicians, mathematicians, and computer specialists understand that binary logical systems do not correctly model all of reality, that most systems fall between the poles of binary understanding and, what is more important, "that [logicians] could formulate logical systems different from Aristotle's" (Coveney and Highfield 1995, 74). In their study *Frontiers of Complexity: The Search for*

Order in a Chaotic World, Peter Coveney and Roger Highfield cite numerous examples of commercial, industrial, and medical applications of fuzzy systems. But the underlying concept of fuzzy logic is a deeply philosophical issue: "Fuzzy classification evidently violates Aristotle's law of the excluded middle, according to which an object either does or does not belong to a given set. But in fuzzy logic, the multivaluedness of truth is actually a point of departure" (Coveney and Highfield 1995, 75).[5]

The story of the sailor and the reflections on the tapestries of Cluny are stories of precisely such complex fuzzy frontiers where exact calculation fails. The dynamics of both involve the passage through skin, primary locus for the tactile sense. The "skin" in the sailor episode is the porthole separating the fiery inside of the ship from the icy sea and ferocious winds outside. It is also the undefinable region where the man's being wavers between subjective consciousness and frozen or seared object-mass. On the sixth tapestry of Cluny, Serres equates the pavilion with the body of the inner sense and the tent canvas with its skin. Both skins function in Serres's reading as frontiers of such extreme complexity that precisely defined and quantified descriptions of their composition become impossible; they are black boxes. He resorts therefore to fuzzy logic to imply their "multivalued" functions.

Adding yet another variable to this complexity, the story of the sixth tapestry depicts the problematic of language, fateful tool for the human as both a sensate being and an intellect, doubly fateful for a poet-fabricator who wishes to relearn the use of the senses by the very tool that distorts their efficacy. Language appears on the tent canvas which doubles for the body's skin; the canvas-skin, site of the tactile, is not only circled by a ribbon of words but also covered with pentecostal tongues. But language also plays a new role in logic: "The key concept in fuzzy logic is a linguistic variable, that is, a variable whose values are words rather than numbers" (Coveney and Highfield 1995, 75).[6] By juxtaposing language, the tactile, and the fine line that separates and weds the two, Serres reveals the process of fuzzy logic with which he will try, not to explain, but to imply what happens in the black box where the senses convert the concrete into abstract forms that our intellect then shapes into knowledge.

Other stories on the tactile also involve some form of skin and converge on the notion of the fuzzy variable which becomes the guiding principle for an understanding of the other senses. The tactile is perhaps the sense through which we experience the object not only at the most immediate level, but also at its most varied, since the sense of touch is itself a conglomerate of variable "skins" or membranes located throughout the body. In the original French version of the fairy tale "Cinderella" as told by the classic storyteller Charles Perrault, the slipper is made of a certain type of fur called *vair* in French, which is homophonous with *verre* meaning glass.[7] There are, in fact, French versions of the fairy tale that speak of a "glass" slipper instead of a "fur slipper." The phonetic confusion gives Serres the opportunity to underscore the difference between the touch of the soft furry skin which allows for innumerable degrees of tactile sensation blending into each other and, on the other hand, the transparent, rigid material of the glass which allows our senses to slide by or through it, to arrive at a certain knowledge of a clear, distinct measure. In other words, the glass slipper is to the fur slipper, what the mathematically or logically manageable quantity is to the variable multiplicity of fuzzy logic and the *mélange* of a philosophical topology. Partly in jest, Serres questions the seriousness of a philosophy that talks about a slipper. (Its French equivalent *pantoufle* adds to the seeming levity by the slight buffoonery inherent in the phonetic composition of this word.) But the epistemologist sees a *nouveau savoir* modeled in a fairy tale about a slipper that variably adjusts to any foot ready for the princely ball (64–65).

The reader has been brought once more to the image of the dance. Described in terms of a floating movement in a topological space of variables, in *Detachment*, and as a first step traced back to the nude foot of the "Unknown Masterpiece" in *Genesis*, this topological model is introduced by the tactile in *Les cinq sens*, where the dance appears in its later chapters under the guise of the Ulyssean method and logic of circumstances. The entire chapter on the tactile is in essence an aesthetic "proof" of the multiple and clarifies why the senses in general would be Serres's favored vehicle for a demonstration of the *mélange*. The skin is the common border where the world and the body touch, where the one who feels mixes with the felt or sensately experienced. It is always variate and contingent, it is a "fuzzy set" which means that it is not a middle or focal point, not a geometric center or point of order, but a pure variability or *mélange*:

> "Mélange" explains this better than "middle." . . . The middle point is abstract, dense, homogeneous, almost stable; it concentrates upon itself. "Mélange" fluctuates. The center point is part of solid geometry as it was once called. "Mélange" favors fusion and moves fluidly (*verse au fluide*). The center point separates, "mélange" softens: the middle as a concept creates classes, but "mélange" produces half-breeds. (82–83)

What is as yet a singular topological explanation of the sensate, closely related to the Serresean distinction between a mechanics of solids and scientific inquiry of fluids and turbulence, is clearly delineated: on one side, the orderly division into linearly defined classes or sets, on the other, the *métis*, Serres's preferred allegorical figure for the mixed and multiple, for nonlinear chaotic *mélange*. For now playing the role of a metaphor, the *métis* takes on a life of his own in *Le Tiers-Instruit* where he becomes the introductory model for a nonlinear dynamical discourse. Today Serres is often called the "philosopher of the *métis*."

Here, the reader must understand, first that the work of the senses always implies fluid experiences and variability; second, that the models Serres presents and of which I have discussed a few at some length all point to the Serresean paradigm of *mélange*, the *métis*. Finally, a solidly scientific discourse is the bedrock of these epistemological reflections on the sensate which, in turn, is identical in Greek with the term "aesthetics." There are frequent reminders that this text, with its subtitle "Philosophy of mixed bodies," is not a metaphorical arena of scientific concepts, but is itself a scientific study by other means, by other models. When Serres speaks of a topology of the tactile, he is not speaking metaphorically, but is using language to express in words *exactly* what computer-generated pictures of nonlinear dynamical systems model visually: fractal basin boundaries which are areas of infinitely complex behavior within a nonlinear system's evolution over time and which appear between regions of stable behavior.[8] Stated in scientific terminology, "[f]ractals display the characteristics of self-similarity, an unending series of motifs within motifs repeated at all length scales" (Coveney and Highfield 1995, 427). The sailor's unresolvable hesitation between subjective conscious and objective being, and the lady's undecidable movement on the threshold of her tent-body are Serres's models of such fractal or fuzzy border zones.

The preceding observations offer a key to understanding why Serres would opt to write of sensate experiences when, in fact, the scientific nature of his reflections is at stake. It involves a profound and epistemologically critical insight that entails a qualitative shift as well: the senses do not merely serve the aesthetic part of our being, separated for too long from the rational part and relegated to the unreliable, fictional, unstable and chaotic makeup of human nature. Neither are they an elegant excuse for an aesthetic discourse on science. For Serres, if (re)lived in the form it should be experienced, the sensate brings together, at incredibly complex boundaries, the subject and the object, our being as an intellect and the reality of the world we are part of. Serres's pursuit in *Les cinq sens* is therefore identical, not merely analogous, to that of scientists. It is an epistemological conflation perhaps best described by James Gleick in a terse, generic summary of the paradigm shift involved:

> Those studying chaotic dynamics discovered that the disorderly behavior of simple systems acted as a *creative* process [emphasis added]. It generated complexity: richly organized patterns, sometimes stable and sometimes unstable, sometimes finite and sometimes infinite, but always with the fascination of living things. (*Chaos*, 43)

For Serres, the sense of touch is the fractal boundary that opens up a creative process, where objective reality and subjective intellect *invent* together. The other senses add to this complexity, as we shall see. It is important to recognize the discursive fractal shapes in *Les cinq sens* in order to understand the complex dynamics of invention at play in *Statues* and especially in *Le Tiers-Instruit* where the métis is Serres's fractal.

The second chapter called "Boxes," a reflection on the sense of hearing, does battle with language on a grand scale, whereas the first chapter is punctuated with local, uniquely semiotic scenes implying a kind of step and counterstep minuet between the word and the tactile. The sense of touch involves local surfaces; sound, however, occupies volume and spreads out globally (149). The sense of hearing is engaged globally to capture sound on three levels, the internal non-linguistic voice of the body, the vibrating silence of the physical world, and the "garbage-noise" of the human collective summarily described by Serres as vile dirt (91–92). The initial scene of this chapter is set in the amphitheater of Epidaurus where the sightseers' strident noise, what the collective calls "language and logic," overwhelms the gentle whispers of a solitary visitor's body. This confrontation is set in the midst of nature's peaceful silence which Kavanagh in his review article aptly names the sound of the "unworded world" (Kavanagh, 938).

The tourist collective surrounds itself with its all-invasive language, as if the latter were a safe ship keeping the unknown and illogical world at bay behind its protective hull (92–93). Once it has left the ruins of Epidaurus and taken its noise with it, the solitary body slowly recuperates its hearing of the world. Serres is implied in the narrative *I* whose visit to Epidaurus is interrupted by an excited, curious (and raucous?) group of tourists. This example stresses to what extent human language, one of the three sources for the sense of hearing, has drowned out the other two. Like the tent-pavilion on the tapestry of Cluny, language has become a safe haven from the chaotic "outside"; the social collective is not willing to leave it. Language substitutes for the other two auditive sources which can no longer compete effectively. As a social entity built on language constructs, we have collectively come to believe that the data of the world reach us exclusively through language. Only when we pay a solitary visit and leave all protective language noise behind does our body begin to rely on its sensate powers, it hears the "unworded" world, it regains equal status with the intellect: there is a healing of the human being. "Guérison à Epidaure" is Serres's title for this fable. Hearing the "unworded" world is an encounter with the gods, the shy immortals that pervade the Greek amphitheater. This is Serres's formulation of the sense of a renewed hearing.

This sensate experience occurs in the same topological space of variables where angels dance with the Chinese farmer and saints accompany the mariner perched high in the crow's nest of the heaving ship, where women join them in an aura of prodigious giving of love (Rilke). At Epidaurus, the Greek gods drift by. Wonderfully harmonizing with Chinese dragon kites, they offer "hearing" that is utterly different from the impoverished noise we call language. As a result, the human intellect can encounter the objective world immediately, without an interfering screen. Deliberately alluding to the Holy Spirit who descended upon the Virgin, Serres's carefully chosen words imply that the fluid space of multiples in which we truly experience the objective world, should also be the space of his culture's god, the same god who has been rationalized by Christianity into a rigid logic, into a concept constructed of dogmatic pronouncements (109). Quietly listening to the whispers of the place, the visitor at Epidaurus encounters the divine as a manifestation of a healing sensate experience. The objects of the world gently approach him, he listens, and in this *mélange* of silence a new language appears (93). Relearning the power of the auditive sense enables us to rediscover the harmony of a language delivered from its strident domination over the unworded world.

But the visit to Epidaurus leaves the epistemologist with the nagging doubt whether, upon parting, the cure has really taken hold, whether the addiction to language will not rear its ugly head again. Therefore Serres embarks on a visit to Socrates, in order to learn if language can truly be anything other than a narcotic power inducing a hallucinatory belief that the objects of the world must come to us through words (96–102). Who is better placed than Socrates, the Ur-ancestor of Western thought, to tell us whether he has found the secret of a language that harmonizes with the other two sources of sound? But Serres finds him incessantly talking on his deathbed; Socrates and his friends are so addicted to the word that they cannot stop chatting. Language encroaches imperiously and absolutely on all immediate experience of the world. The dying philosopher is the orchestrator of this gigantic language spectacle which, Serres reminds his reader, we piously call *Dialogues*. It seems that only death can bring a cure from the addiction to representation by words and signs so powerful that Socrates' prison of stone from where he refuses to escape is identical to a fortress of words, to an ivory tower formed by the logical rules of his close-knit "big boys" club (96–97).[9]

With Socrates as the master of this founding scenario, Serres mockingly wonders why "twenty-five centuries of philosophy weep and mourn [the sacrifice of] this ugly looking exhibitionist" and then, discouraged, he leaves the theater at Epidaurus (96).[10] He takes stock of our chances of ever sensately experiencing the world before language builds up its ramparts around us. Surrounded by its single-minded logic, philosophy sets up house in the theater abandoned by the fragile divine (101).

If Socrates himself experiences the world only through words, we must sadly conclude with Serres that we, too, cannot do otherwise. We do not heed our god who, it is told, "'lived among us as the "Word made Flesh" (Jn 1:14), that is, made to become one with the materiality of the world. Instead, we have cloned him into innumerable sermons and books, into a maliciously festering cancer of words. We drown out the divine with our tourist cacophony and logistical data gathering, like true disciples of loquacious Socrates who was delivered from both prison and language only by death (101). We are his descendants and remain mired in our stagnant cravings for representations. Serres hopes to change the stakes; in the ruins of Epidaurus he sensed the immortal vitality of reality as an alternative to death. In a language whose powerful terms seem to vault from the pages (108–09), Serres tells his reader that this vitality cannot spring from our intellect and its representational expressions alone. It must also derive from the willingness to accept an empirical truth. The sensate is the only tool available to us for approaching the empirical unencumbered. The "unworded" use of the senses opens the door to this new and yet so ancient knowledge.

Turning his back on the ugly countenance of the talkative philosopher, Serres searches for true hearing in the realm of music as the aesthetic link between raw noise and language. Offering music as a "good" source for wise hearing is, however, not a simple gesture in Serres's text, in the sense of Kavanagh's cautious advice not to expect a treatise on the five senses "in any simple way." After all, Serres could then in good faith end his discourse and perhaps invite the reader to spend the rest of his evening listening to Charles-Camille Saint-Saëns's *Rondo Capriccioso*. But such a solution would leave unsolved the question of language and its powerful screening agency so fateful for the human inventor and user of language. Instead, Serres turns once again to myth, as he had done in his implications on the tapestries of Cluny, reconciling the lady's sensate experience of the objects in her garden with language in the form of myth (the unicorn). Myth is "a use of language which nonetheless refuses language's drive to total hegemony" (Kavanagh, 939).

In a lengthy passage on various aspects of the myth of Orpheus, Serres finds a way to make us hear language through the power of hearing music. This section of the second chapter (137–46) is perhaps the most intricate in its implied logic, of the entire text of *Les cinq sens*. When Eurydice dies from a viper's bite and enters the netherworld of shadows, all that is left to Orpheus is her name. More importantly and since we are reading a myth, she has always been a name and a concept only, whereas, epistemologically speaking, Orpheus is the activator of the myth and of the cult that carries his name. What Serres wants us to hear, though, is not the song of Orpheus' lyre, in the conventional sense of listening to musical renditions. He offers an example of a musical language which effects the opposite of our logical, imperious language. The latter is the screen of representations, linguistic signs, and concepts, converting the physical world from the "concrete" or hard (*le dur*) to the soft or immaterial state (*le doux*). In contrast, Orpheus' musical language or language-music changes the "soft," the shadowy image, or a word like "Eurydice," into the "concrete," the real, the "hard" and tangible, the beloved woman in the flesh.

Serres tells the Orphic myth in a way that makes us hear, not the lyre's music gently coaxing Eurydice from the shadows of Hades back into the world of the living, but the myth itself that tells the story from a different perspective: how a musician uses his enthralling art, so powerful that it charmed the Furies themselves and drove them to tears, in order to change a shadowy name to a woman, to a creature of flesh and bones. If we use our sense of hearing to listen to the myth, we will experience the magic conversion from the linguistic sign to the concrete, physical object, a totally illogical process by all norms of representational thinking. Our hearing must go beyond Orpheus' music that charmed beasts, rocks, and stones, even mighty Pluto. Attuned to the music of the concrete, we experience sensately and accept the slow, piece by piece, and step-by-step "fabrication" of the flesh-and-bones human being from a linguistic abstraction (138–39).

Hearing, which aims to fabricate life, is not easy. The slow ascent from the netherworld is arduous. Serres compares it to a long and patient voyage along the complex corridors of a labyrinth (140). Behind Orpheus is nothing but a shadowy phantom and death; in front of him a being is reborn, cell by concrete "hard" cell, with each note he coaxes from his lyre: "The word becomes flesh. . . . Orpheus invokes her name, his [living] voice and the lyre's cords vibrate, he calls her, shouts, sings, and chants in rapture. *He composes music and Eurydice*" [emphasis added] (139). Reading the Orphic myth as Serres reads it, we begin to climb out of our prison-maze of language toward the real world in order to partake of the feast of objects, just as the lady on the sixth tapestry of Cluny steps out of her tent-pavilion of language to enjoy the objects of her garden. We are close to reconciling our subjective intellect with the objective world around us, through our sense of hearing.

Orpheus' task is the most difficult imaginable, writes Serres who does not refer to the taming of the wild beasts, the appeasement of Pluto by his charmed music, or the reduction of objects to their representative linguistic or musical signs. The most difficult is the inverse, changing the representation of an object into the object itself (140). Orpheus himself fails at this task and forever loses the beloved woman. Resembling a most difficult but wonderfully human birth, it remains an impossible task (141). I can only urge the reader to read the entire Orpheus passage once more and to listen to the many interstices of this truly amazing discourse. But Serres knows that he, too, fails in his attempt to create a body with words, just as surely as did Orpheus. Indicating the way out of the shadowy world of abstractions, as does the Orphic myth, he now needs his reader to change the gentle invocation of a name into the beauty of the concrete object. In other words, the reader is asked to carry on the musician's task, through her reading and at precisely that moment of creation when Orpheus turns his back on it and, overcome by his longing, is lured once again into the netherwordly realm of words, signs, and representations.

The Bacchantian women who tear Orpheus apart and fling the many pieces of his body throughout the countryside merely follow Orpheus' example. Orpheus looked back to verify (analyze and understand) that his beloved was really there, and the women take apart and analyze in order to understand his mysteries. He stopped composing and thus caused her emerging body to decompose once again into a name and a shadow; the Bacchantes, likewise, do not understand the composition of the concrete, they pick it apart and thereby decompose the body.

If we truly hear the myth, we do otherwise and recompose, we pick up at the point where Orpheus failed, and we begin to partake of a miracle: the conversion from the name, from the "soft," to the body, to the "concrete" (143–45).

The dispersion of Orpheus' body is the mythic expression of sound as a distribution in volume, in contrast to the object of touch which is experienced locally on the "plane" of the skin. Hearing Orpheus' music as bodily dispersion is another step by which Serres proposes the sensate experience of the world, shifted to global dimensions but not necessarily to more precise information. He describes the dispersion as an effort to construct small black boxes, or "listening devices" attached to the original black box. But these little black boxes themselves mimic the space of input and output, fashioned with their own set of walls. Needed are yet other boxes or listening devices to receive the information processed through this second set of boxes, and so on ad infinitum (146–48).[11]

The chain of boxes forms an open-ended loop, because each box requires yet another inventive device to translate its emission. This kind of dissipative listening is compared to a dance and to a nonlinear movement. Our journey to a genuine hearing resembles Orpheus' labyrinthine ascent. The myth finds no genuine solution for this journey, for it tells us that the easy way out is what Serres calls the "Cartesian method" which looks for the most direct, most linear, and shortest route and presents us with death or the shadowy linguistic sign, whichever result we wish to read into Orpheus' impatient look over his shoulders (150–51).

Appropriately enough, our own body is the most intricate and labyrinthine box of assembled boxes. In Serres's presentation it becomes a labyrinthine starting point, from where we must follow the dance of "topological" hearing, even if the "end" is never in sight. This is exactly what we do, maintains Serres, when we construct our houses, our social organizations, our epistēmē, our languages, and our abstractions. All are black boxes, each designed to listen to its predecessor in a long chain of hearing; "we multiply our arts and strategies in order to avoid Orpheus' gruesome destiny" (154). But surrounded by so many walls, we forget how dangerous and cruel the outside is; after all, the myth compares it to a long passage through hell. We let down our guard and now and then think we have arrived at the exit from the labyrinth, found the solution, and now hear the world. Then we resemble the philosopher who, sitting at his window and looking at an apple tree, puts down on paper a lengthy treatise on the concept "tree." He has turned the "hard" into the "soft" mode of representation and words. We fall back into the shadowy soft world of signs. Our epistemological efforts to escape from the labyrinth create representations of the world which we mistake for the real one.

For Serres the epistemologist, these efforts focus again on the mystery or fuzzy zone of the black box. In this process, our epistemological and social endeavors always end up creating yet other black boxes in order to understand the screening process of input and output of each previous box. For Serres the writer, language becomes, of course, the ultimate screen of abstraction, the black box interposed between the sensate experience of the physical world and the subjective intellect. But as if to reinforce the irony of his position as a "user" and "fabricator" of words to convey this message and to counteract the effects of soft language, Serres offers his reader a linguistic concretization of the labyrinth of epistemological boxes we have created: He *writes a labyrinth*, beginning a sentence on page 157 that he does not close until page 159—a sentential maze of fifty-eight lines containing in excess of six hundred words. The sentence meanders from strings of adjectives to multiple participial constructions, from subordinating to coordinating clauses, from conjunctive to relative clauses, held together by no less than ninety-six commas (157–159). Following the lengthy discussion of the Orphic myth, this syntactical monster is clearly meant to represent the torturous, serpentine climb from the shadows of the netherworld to the clarity of the real world, to the union of subject and object. It tells the story of the sum of all human listening, through the ear of the social collective as well as that of the individual body. It tells therefore the tale of the box-upon-box construction, how it allows the sounds of the world to filter in, now and then, how it always screens them, and how this mighty construction is traversed by the crowning glory of language that is the ultimate box for the overall output called "meaning and knowledge."

The myth tells, however, that Orpheus did not make it to the last level; somewhere near the mouth of the long upward spiral he turns around in the mistaken belief that he will see and recognize (know) the beloved object. As soon as Serres has completed his labyrinthine sentence, he changes its direction and calls it a "spiral descent into the shadows" (159). What I can indicate here only in a linear fashion is in Serres's text, however, a spiraling movement that has no single direction but is an expression of our open-ended efforts to listen. It indicates our fate by which we are always tempted to peek over our shoulders, thinking we have grasped the object. The single meandering sentence (157–59) is the story of our hearing as a replica of Orpheus' story, but magnified to Sisyphean proportions.

With the kind of hearing Serres proposes throughout this chapter, I urge the reader to pass through the labyrinthine sentence. It reveals Serres at his most poetic and dramatic. Its expressive power is such that it conjures up the portrait of the human being whose intellect, separated from herself as a sensate being, withdraws behind ever higher and more intimate walls of her own making to protect from what is perceived as an alien world. But the poetic and fabricating power of the sentence is such that reading it has a cathartic effect: it opens a meta-level from which we can perceive the open-endedness of our box and wall constructing. Like the myth which represents "a use of language which nonetheless refuses language's drive to total hegemony" (Kavanagh, 939), the noise of the sentence suddenly opens our ear. It enables us to begin Hearing. For an instant, the duration of six hundred words and fifty-eight lines, language becomes opaque, hard and concrete; it becomes a physical object and we hear the sound of the physical world in this exemplary object. The sentence screams our weakness: that we keep turning around toward the shadowy world of signs. But its poetic sound is also the scream of a rebirthing and the hallelujah of a renaissance (381). We *begin* to hear sensately.

The third chapter of *Les cinq sens* concerns the senses of taste and smell and opens with what is best described as a discursive wine tasting party. A glass of Château Yquem 1947 serves as the imaginative key with which Serres launches his elegant arguments for the synaesthetic sense of the gustative and olfactory (165–66). The two are linked, not for convenience, but essentially, because both what is tasted as the wine passes over the tongue and what is sensed when its bouquet reaches the nose constitute the one unique sensate experience of an old Yquem. Yet this quintessentially empirical event cannot be analyzed as to its characteristics of tastes and smells, because the effects of the combined senses form an "unstable shimmering material, a mixed body (*corps mêlé*)" (169).

We have arrived at the heart of "Les cinq sens: Philosophie des corps mêlés." The wine's bouquet and taste are the resultant of an inextricable mixture of climatic conditions stretching far back into the past, of ancient soil formations and changes, of many winds and sunny periods, of cultivation honed and refined over centuries, of choice of cave, barrel, cork and bottle. Whenever, genuinely, the nostrils smell or the tongue tastes a fruit, a berry, an herb, a flower, the senses are guided by a sort of road map of chaotic sensations. The maps Serres speaks of are not bottle labels (*étiquettes*, in French), poor substitutes offered by language, the "other" gift of the mouth that analyzes the gustatory and olfactory sensations and brings order into their chaotic mixture. The maps of *mélange* are the direct output of the two senses, beyond and before any etiquette or language: a pure landscape of sensations which never ceases to shimmer and palpitate in an unstable way. In other words, the sensate map undergoes continuous changes.

Actually, Serres is implying a topology of mappings, thereby echoing his own tale of the topological space in which saints, angels, lovers, sailors, and Chinese farmers dance. This space implies a unique expression for a true understanding of time. Similarly, the wine's taste and smell assemble and combine in a non-integrable formula the many spatial and temporal conditions necessary to produce the wine, of that year, of that particular picking. Therefore, in the midst of his discussion, Serres takes stock and simply states: "I am drawing a thousand maps; I am but speaking of time" (172).

The exquisite study of the senses continues the quest for a genuine understanding of ahistorical time. Called a "new knowledge" in *Detachment*, this quest now becomes a new wisdom which is nothing less than the original and ancient identity of sapience and sapidity. Serres bases their identity on the similarity between the two words, and on the fact that both taste and the pronouncement of wisdom pass through the mouth and use the same organ, the tongue (167).[12] He maintains that language, even as a vehicle of knowledge, is not the same as wisdom. We forget too quickly, he continues, that homo sapiens was initially the twin of a being who tasted, not of a being with words (167).

The Serresean wine party ends with the wish that we might have not only sapience but also sagacity, the fine quality of discernment which comes from a keenness of sense perception. "All banquets should be held under the motto: sapience and sagacity; then only wise tongues would be present around the table" (177). The "tongue" is here understood in a multiple sense that the reader should not try to dissect into language, tongue, and/or word. Suffice it to say that Serres speaks of three tongues, the tongue of taste, of smell, and of language, thereby ranking human sensate power equal to his intellect. Only after having established this kinship does he proceed to tell the story of the usurpation by language, and how the olfactory in particular has been relegated to a baser level of our instinctual being, even below the gustatory. For, more so than taste even, odors seldom remain pure multiples. The most sublime perfume borders on and mixes with the stench of decaying substances. The richest odors emanate from the dead and the putrefaction of their bodies; they become the soul leaving the deceased body, by what is aptly called in both French and English "the odor of sanctity" (178). Since the line between the sacred and deadly decay blurs in the exercise of sagacity, epistemology tries to efface the latter. It strives to sever our links to a wisdom acquired through the senses which would bring us closer to an empirical knowledge of death. From the ritual practices of ancient priests to the work of modern scientists, knowledge acquired through the abstractions of language utterly ignores any experience gained through the olfactory (179).

Serres's interest in the sense of smell and its benign (and not so benign) neglect throughout the ages of rational thinking is not an isolated case or only indigenous to his particular epistemological discourse on an empirical rapprochement with the human intellect. The year that *Les cinq sens* appeared saw the publication of a German novel which became an immediate best-seller, *Perfume: The Story of a Murderer*, by Patrick Süskind. Its narrative intrigue is centered on the forgotten story of a Frenchman who lacks all bodily odors and therefore is marked as an outcast. But he has an absolute sense of smell; his quest for the most sublime odors is a quest for the meaning of his life. Ultimately it is a mystical journey to find himself, an eighteenth-century character whose composite portrait is focused exclusively on the olfactory to which even his superior intellectual faculties are subjugated. Both the novel and Serres's commentary on the sense of scent share the persistent idea that the olfactory is a basis for acquisition of knowledge, this in the face of a strong and successful philosophical tradition of favoring the sense of sight and to a lesser degree the auditive as universal sources of reference (*Les cinq sens*, 184). The two texts arrive at diametrically opposed results, Serres's passage leading to a veritable "Magnificat" for an empiricism which enables us to live time ahistorically, while Süskind's narrative shows that even the "lowest" sensate domain can be an all-powerful catalyst for the emergence of a genius, albeit one of evil and abomination. But both texts contribute to a new epistemological discourse on the sense of the olfactory.

In his novel, Süskind demonstrates how the lowest of the senses impacts on human lives and history. This flies in the face of an enlightened culture and its rational discourse that completely erases "a domain that leaves no traces in history: . . . the fleeting realm of scent" (Süskind, 3). For the Serresean reader, however, the real interest in recalling Süskind's work is summarized in Richard Gray's characterization of the novel's objective, namely, to achieve

> the reconstruction of a history that has left no vestiges whatsoever in official history. Insofar as the novel is concerned with the recollection and documentation of what has been wiped off the slate of historical memory, it pursues a project of critical anamnesis. It is no coincidence, moreover, that this counter-history relies specifically on the documentation of olfactory data, since it is a commonplace of modern osphresiology that of all the senses, olfaction is the one most intimately connected to the function of memory. (Gray, 491)

In Süskind's novel, the sense of scent is a tool for a fictionalized critique of epistemic gestures of exclusion, practiced during the Age of Enlightenment when a-rational spheres of human activity were relegated to marginal domains which lost therefore all title to historical recall. The objective of the novel is above all a global correction of historical time, from a Girardian perspective. By revealing the exclusion of all that enlightened reasoning cannot readily cope with, the novelist renders the willful loss of memory ineffective and thus restores a genuine sense of history. That the latter is presented in the form of a most heinous and monstrous criminal activity makes both, for a compelling call for historical integrity even in the face of human evil unmasked, and for the literary success of the novel.

Serres's meditation on the olfactory plays a similar role in the context of his critical attempt to redirect our sense of time. However, there are two aspects that are essentially different from Süskind's project. Serres's objective is now detached from all accounting of the murderous clamor of history, he is now drawing the topological map of a "new" time. Second, the osphresiological connection of olfaction and memory is more than a narrative strategy, even if it is a strikingly persuasive one as in Süskind's work. It is a necessary step in Serres's argument to proceed from the olfactory sense to memory in order to convincingly imply the relationship between the dynamical time he is slowly extracting from his fables and the sensately knowing existence in the immediacy of concrete reality.

In the passage entitled "Memory" Serres deploys a tightly argued chain of perspectives on the sense of smell and arrives at yet another way of perceiving time, knowledge, and topological dimensions of human space (183–205). The emphasis is on the relatively "unstable" nature of any odor, because it relies on air and its non-analyzable variations as its carrier (183–84). Odor is sensed in singular, ever variegating traces. The more fleeting an odor, the more intense is the recall of a moment or an object of the past associated with its particular composition of confused, mixed, and intersecting elements. It is accompanied by "memories in overabundance that surge forth ecstatically at the edge [of knowledge]" (185). These multiple descriptions of olfactory experience conjure up the delicate aroma wafting from a madeleine cookie steeped in tea, on which hinges Marcel Proust's gigantic effort to create a lifetime by meandering through the dynamical space of remembrances. Serres gives no hint of his argument's possible kinship with Proust's celebrated anecdote. Yet, we come to understand that, once the chaotic character of olfactory effects has been recognized as the twin of the twisting and confused realm of a thousand delicate threads of memory, Serres's text, too, arrives at its "hinge" or focal point. It is the stunning presentation of an apparent conceptual contradiction that I shall refer to in the following chapter as a dynamical *stance*. The subtitle of *Les cinq sens* calls this evocative nexus a "philosophy of mixed bodies"; within the text, however, it is recognized in the expression "circonstance."

The "circumstantial" becomes the metaphor par excellence of *Les cinq sens*. Born from new insights into the power of the senses, and in particular olfaction, the discourse of circumstances marks the rest of this text and links the human as a sensate being to the notion of ahistorical time that *Detachment* proposes as a dynamical space for the harmonious union of subject and object. The Serresean *mélange* of odor and memory is an emanation (*une vapeur*) which resembles, more than anything else, a "circumstance" (184). The use of this term is a fine example of the author's manipulation of metaphors in a radically literal fashion which breaks down their hackneyed associations and enables him to use them in ways that are strikingly, sometimes mysteriously, often monstrously, inventive. This is no idle play, of course, for in *Le Tiers-Instruit*, invention becomes the leitmotif for Serres's agenda of a new knowledge. But for now "circumstance" is an ideal term whose economy of expression sums up the Serresean key notions of *Les cinq sens* leading to a dynamical concept of time.

What I have characterized in chapter 1 as the locus for a conflation, of regularity in the *-stance* with the topology of possibles in the *circum-*, is now presented on the level of the sensate and acquires increasingly the characteristic features and vocabulary of nonlinear dynamical systems theory. The privileged senses of sight and hearing, writes Serres, produce a memory of invariants or stable variations; this

> memory dresses up as knowledge, and its repetition becomes habit that changes soon into law. But the rare trace in the vaporous air, the unstable and

> complex mixture, or the half-loosened knot trailing a thousand threads do not reduce to or arrive at any state of invariance: they are too circumstantial to do battle; rather, they are rhythmic, unpredictably strewn and diluted; they are chaotic. Knowledge, on the contrary, eliminates these unstable circumstances and levels what stands out as rare. Knowledge stipulates: Under the same circumstances . . . (185; ellipsis is Serres's)

By gently taking exception to the important and necessary condition for the Aristotelian principle of identity (under the same circumstances), Serres indicates how far his discourse is removed from exclusively linear thought, even to a point where the logical equation $A = A$ becomes one singular stance or rarity in the topological space of circumstances. The linkage, of the olfactory sense with the Serresean notion of circum-stance, produces the osphresiological concept of a memory born of an olfactory singularity or rarity (the space of aleatory turbulences in the air that carries an odor) in the midst of an "overabundance" of remembered circumstances.[13]

> This memory does not change into knowledge by simple repetition on the lowest denominator. . . . Rather, as confused sense of unpredictable encounters, and as rare sense of singularities, the sense of scent glides (*glisse*) from knowledge to memory and from space to time; surely also from things to beings. (185)

Circumstance is not merely a trope with which Serres now replaces the black box as a descriptive approximation of the dynamical process underlying all true sensate experience. The word becomes itself a literal object whose implications form the passages between the senses and the various Serresean notions expressing the union of subject and object in the topological space of ahistorical time. Circumstance is a bouquet of intersecting sensations constantly in turbulent motion which cannot be quantified or analyzed. At other moments, Serres calls this a "non-integrable system."[14]

A good example of circumstance is Serres's empiricist perspective concerning the core concept of Judeo-Christian tradition: the Word made flesh. A banquet of bread and wine is now recalled. Unlike Plato's banquet of words or Socrates' feast of dialogues, the Last Supper commemorates the Word made flesh. But the com-memoration has become a mechanical repetition where the divine is confined to rituals of language. Serres compares it to Plato's banquet whose statuesque divine specters reduce everything to words and only predicate death. The statue as the paradigm of *stasis*, stagnancy, robotism, harbinger of death, of all that for Serres is diametrically opposed to dynamical thinking, is here situated at the heart of what a long tradition has made of the Last Supper: the feast of stable and invariant Peter the Rock (192–97).

But we have forgotten the other feast celebrated at the Last Supper; Serres calls it the feast of John, the apostle of love.[15] Here bread and wine are tasted and smelled. In the course of this feast, objects of our sensate experience are apotheosized. Divine mystery, in turn, changes into objects whose materiality we share intimately in our own bodies. Based on the sharing of bread and wine, the feast of John is one of sensate variabilities and multiples. Süskind conjures up the fleeting realm of the olfactory against all ratio-historical forces, but can do so effectively only in a fictionalized setting. Similarly but more boldly, Serres invokes the last supper of empirical delights, the feast of love. He does so against two thousand years of ritual that re-*inscribe* the Word-made-flesh into a stable construct so invariant, that together with Hellenistic thought, it forms the *rock* upon which all of Western knowledge is built, leading to today's technological abstractions and virtual realities. With its chaotic variabilities of tastes and odors, however, the banquet of John resembles in every respect the feast of the sixth tapestry of Cluny at the moment when the lady steps out of her tent of language. It resembles the love celebration glimpsed by Rilke in profound contemplation before the same woven tableaus. But in contrast to Süskind's pessimistic review of enlightened rationalism, Serres is on a quest to reconcile the stable and the variable:

> I dream of writing a third banquet where the statue, the angel of revenge, would accept the invitation to drink a fine wine with the one who has conquered his urge to best his guest (194).[16]

Therefore Serres writes on circum-stances, a book on the feast of variables (circum-) wedded to the feast of the statue (-stances), in an effort to stop what has become our deadly history here allegorized in the story of Dom Juan who is condemned by a statue. The same fate befalls John the Evangelist whose message of love is marginalized in favor of the mission charged to Peter the Rock. Both examples are parables of the negligence and destruction of variable time. *Les cinq sens* is a text on circumstances. It lays the foundation for *Statues* in which Serres discovers the statue itself (as *stasis*) to be a dynamical space in which human beings can truly experience time as life-giving and no longer as the mortal combat zone of our quarrelsome relations. For now, however, the discourse on the five senses tries the simple gesture of getting the two, the invariant and the nonlinear variable, to sit down *together* at the banquet table.

Under the aegis of the circum-stance, Serres's discourse can now logically digress from the senses to expose what has driven an apparently unbridgeable gap between empirical wisdom and mediated knowledge founded on language or, allegorically expressed, between love and mere talk of love (211). The Serresean statue as stasis or *-stance* models "talk" as that form of linear, rational knowing which becomes more and more rigid, unbending, where even human beings as bearers of language begin to resemble the robots they create (211). The *circum-*, on the other hand, represents the topological space of multiples and variables, all that cannot be harnessed rationally. By presenting the sensate and in particular its stepchildren, taste and smell, as the paradigm for this undefinable domain, Serres offers an example which we can recuperate as part of our own immediate, corporeal makeup, even though our intellect has for so long carried the day.

Passages with the subtitles "Statues" and "Death" (205–39) expand on the historical drama of this struggle between stable, rational, linear thought and the "fleeting realm" of empirical wisdom, but not before Serres has summarized the vital stakes involved: In today's highly advanced technology which submerges us in a mediated world of codified formulas and byte-sized information, going far beyond ordinary language, empirical wisdom has become superfluous. Philosophy returns to its true calling only if it becomes once again "love of wisdom"—and of sensate wisdom. Serres does not reject rational thought and the work of the intellect with its avowed aim of bringing order into the empirical world (214). Rather, he wishes to "seduce" the -stance so that it may let go of its competitive urge always to get the better of the senses; he proposes the banquet of a true symbiosis, *the feast of circum-stances*. The lengthy discussion on the Word-made-flesh with its accompanying imagery of biblical scenes concerns the obstacles that have to be overcome: By a long tradition culminating in our technocracies, the Word-made-flesh has been mediated into "the sensate-made-word" and today is the rationalist monolith of logic and scientific knowledge. It has become Peter the Rock, or the Banquet of Plato's dialogues, the stable and invariable world of our intellect, to which we erect statues of all kinds in acclamation, and which Serres sums up in *the statue as such* as the ultimate acknowledgment of stasis.

The Word was made flesh through a woman said to be "full of grace." We are mostly concerned with her state of being, for the child conceived of the Holy Spirit was destined to change the course of history. But Serres perceives the woman full of grace *before* the Spirit of God descended upon her. Before, she is full of grace; after, she is filled with the Word. The divine word overshadows everything, absolutely, fills her and us absolutely, and overrules what cannot be rendered into words. But before all this,

> before the Word comes to her and she conceives, before language and concept, before even her immaculate virginity required by the Word and created by It, she of flesh, she the mother, she the woman, she of physical sensitivity, lived full of the [immediately] *given*. (219)

For Serres, the woman full of grace is the paradigm of unmediated reality, she lives the fullness of the immediate world, she incorporates the condition under which the divine Word can come into this world and become one of us, the "Son of men." That is why we speak of a virginal conception, writes Serres for whom the virginal means that state of fullness of grace that no word has yet penetrated and separated from the immediate world freely given. But the "Word-made-flesh and made-words-again" conspires with Socrates' endless talk to produce our cultured blindness which cannot recognize the fullness of grace, because we have fallen from grace, from the unaccountable state of variabilities that the sensate alone can approach. Sadly, Serres admits that words are our philosophy and our religion (220).

The -stance develops from the Word, Judeo-Christian and Hellenic, into an all-encompassing system of philosophical, religious, and scientific dictates (221–22). In the final analysis, we have erected a colossal statue in honor of an epistēmē based on an opaque screen of language and an absolute logic, a deterministic monolith. This is implied in the many statue-like phenomena Serres discusses throughout his writings; together they produce a haunting echo that finally reverberates into a reiterative dynamism in *Statues*. By contrast, the "circum-" or nonlinear multiple aspects of our sensate experience of the world are embodied in the woman who does not understand and is left speechless when the angel comes to her (220). Long forgotten, because gratuitous fullness (*gratuité*) has given way to the onslaught of language, the woman is not present at the feast of the Word/Rock/Commander (231). But she is not alone. Mary-full-of-grace is merely the most exalted among the women who remain in the background at the Last Supper. Instead of talking, they act, whether it is preparing the feast of taste and smell, gratuitously in the background, gleaning the wheat fields of Boaz with Ruth (223–24), or hurrying to anoint Christ's body in the grave. In the simple presence of grace-fullness Serres attributes to these women, they are soul-sisters to Rilke's women-in-love. For both the poet and the epistemologist, the woman of the forgotten feast of love is the embodiment of the circum-, of non-integrable and non-analyzable variability.

Although Serres often refers to what may summarily be called "the feminine," his writings do not betray feminist tendencies in the accepted sense of the term as it applies to contemporary feminist theory and writing. In *Les cinq sens*, the feminine relates, first of all, to a particular style of thinking that is gender-less *because* its language is gendered. This argument does little to convince dedicated feminists whose efforts to degender language, especially in the sciences, have produced a remarkable awareness of the pervasive role of gender-laden analogy and metaphoricity in scientific and other discourses. But for a French philosopher teaching at an English-speaking university and for whom language and style are an integral part of the content of his message, the argument is eminently valid in light of the differences between the English and French languages. In an interview, in which he speaks as a member of the *Académie française* whose task today is the conservation and proper use of the French language as it evolves, Serres notes the English forms of *a* and *it*, which the French renders with a gendered *un* or *une* and *il* or *elle*. These are examples of linguistic gender distinctions which are so deeply embedded in the very foundations of French speech and thought that "the problem of gender is for us less related to the sexes than in English" [author's translation] (Giusti, 113). The problem of gendered distinctions whose origins are traceable to biased analogies or metaphors are, of course, not addressed by such a response, as Serres himself reluctantly admits on the insistence of his interviewer. Nor are the views of philosophers and authors such as Simone de Beauvoir or Luce Irigaray whose language is French and who have variously addressed the burdens placed on the feminine by the biased culture they live in. In Serres's text, we may want to understand the feminine as an expression more akin to Goethe's *das Ewig-Weibliche* which, incidentally, presents its signified through a gender-neutral signifier.

Much more significant for our reading is another function the feminine fulfills in Serres's discourse. In *Genesis* he accepts traditional interpretations of the woman as "other" in the guise of the chaotic and nonsensical element in Balzac's "The Unknown Masterpiece." Chapter 1 of the present study examines the consequence of this uncritical acceptance for a text that carries a "new" knowledge as its stated objective. Conceptually, argues Genevieve Lloyd in *The Man of Reason: "Male" and "Female" in Western Philosophy* (1984), the Other appears at the moment when the Self transcends its biological destiny and becomes a moral being of values, a rational being of logic, and a being of technology who uses tools to shape and control the world. But since woman in her maternal being is and remains bound to her biological body, she cannot raise herself completely above her animal-like destiny. Part of her remains part of the world which is now the other for the transcendental being. But since the woman is very close and similar to man, he must distance himself from her with particular care; she becomes for him the quintessential other. "'Transcendence', in its origins, is a transcendence of the feminine. In its Hegelian version, this is a matter of breaking away from the nether world of women" (Lloyd, 101).

Since Serres's objective in *Les cinq sens* is the undoing of the hegemony of precisely such a transcendence and the rediscovery of the values of bodily experiences for the human being, the feminine as other is a preferred metaphor. It is the perfect allegorical introduction to the philosophy of the *métis* and prepares the later discussion of the much more generic "tiers-instruit." The Other is no longer the outcast but becomes valid on its own (chaotic) terms as "other." In the figure of Mary as the absent woman, therefore, Serres undoes Christianity's near-apotheosis of the transcendent human in the Virgin Mary, because he needs to recall for us the forgotten role of the "other" woman, the one full-of-grace. By virtue of her immediacy with the *given*, she shows us what it means to live and be without language, without the word, what it means to be a derelict subject living in the sensate fullness of the world's objects, before language and before the word. In this revealing glimpse of the Other in the figure of the "forgotten" woman, Serres paints a first portrait of the "new subject" who invents independently of the sacred ivory towers of science and knowledge, and whose actions start from a secular (*laïc*) viewpoint, excluded from the inner (transcendent) sanctum of the initiated (370).

Laïcité as used by Serres, especially in the final chapter of *Les cinq sens*, goes far beyond the historical importance of French laicism in the nineteenth and twentieth century, attenuated in recent years with the rising interest in noninstitutional religious expression, but sociopolitically still reverberating in the often bitter and passionate debates on school reforms in France. Unless the reader is familiar with the specifically French phenomenon of laicism, the fullest significance of this term for Serres's discourse will escape him. In this context, it does not denote the laity as a group of faithful which together with its clergy composes a given religious community. Altogether other when used in the philosophical and historical-political sense the French attribute to it, laicism counters religious dogma, in particular Catholic dogma in historically Catholic France. It is a philosophical position of rational thought for the body politic, it is the accepted conception of the world by a republican state without an official religion. Embracing laicism meant that post-revolutionary France evolved into a secular state with its own rationalist metaphysical principles (Curtius, 178–80).

Of course, Serres plays on these historically meaningful connotations the concept of laicism holds for his French reader. But he does so only to accentuate, for that same reader, the deeper meaning of *laïcité*, namely, to renounce the original meaning of the sacred as a space (the *templum*) which excluded the uninitiated and thereby gave rise to the dichotomy of the excluded/included, or the sacred/profane. Serres's laic perspective is his first premise for the philosophy of circumstances which develops into the philosophy of the instructed-middle, where inclusion and exclusion no longer play a meaningful role. Significantly, his later study by the same name, *Le Tiers-Instruit*, begins with a prologue entitled "Laïcité."

Paradoxically, Serres needs to recall the "word-less" feast of gratuity, *because* language, too, is today threatened with exclusion by oblivion. He recalls it through the woman who gracefully gave herself so that the Word could become flesh, because the word as such, systematically multiplied into our language-dominated world of signs, is itself dying under the weight of simulated models, algorithms, bytes, and virtual reality. Therefore, Serres conjures up the woman, not of the Annunciation, but the one "before," full of sensate experience, the one who is ignorant of the exclusive sanctum of words, *the laic one* (233). She alone can come to the rescue, she alone will gracefully donate all her being once more, so that language can be made flesh once again and then will grow into words that will no longer lay claim to any hegemony over the world of objects.[17] Again appealing to mythological language, as he had done when reflecting on the *Lady with the Unicorn*, or when writing on the Orphic myth, Serres hopes for a renewed and grace-ful birth of the word. So he salutes the woman-full-of-grace. His discourse is interspersed with the angel's Ave, which imperceptibly becomes his own Ave to the state of grace (217–254). Rewriting the myth, Serres becomes a modern Gabriel who announces the renewal of the Word and assists at the rebirth of a genuine sensate experience that had originally been subjugated by language.[18] By writing on the aesthetic powers of the human, Serres acts as midwife to a rebirth out of a state of grace-fullness. What will be born, he hopes, is a new knowledge and a new wisdom.

Announcing the word-in-aesthetic-fullness is precisely what the notion of "circumstance" does. It reconciles linear stability with non-integrable variability; or, reduced to simplest terms, order with chaos. Ten years earlier, Serres would speak of a dialectics of circumstance as "the only exact dialectics," much truer to reality than the Hegelian *Aufhebung* itself (*La naissance de la physique*, 78). Now Serres can simply conjure up the contents of a bottle of Château Yquem and, by finding it virtually nonanalyzable, pronounce a philosophy of circumstances which itself becomes a loyal part of the reality it wishes so ardently to describe (240). The empiric remains a mysterious black box only to the scientist who, when faced with something like "a singular non-integrable essence," speaks of fractal reality (241).

Entitled "Visit," the fourth chapter of *Les cinq sens* finally deals with sight, the most privileged of the senses. Adopted by early philosophical language as a metaphor to express concepts, sight became a metaphysical premise in its own right: "idea" derives from the Greek *idein* meaning *to see*. For Serres, this is exactly what is wrong with our habitual seeing which, honed by a long tradition, provokes us to abuse this powerful sense, in a wholesale fashion. The privileged position of sight comes with a high price; our philosophical tradition uses seeing to formulate ideas, to mediate our experiences with the world, to summarize concepts into sweeping, generalizing theories. The term Weltanschauung demonstrates dramatically how language has mediated an empirical viewing or seeing (*schauen*) of the world (*Welt*) into a comprehensive conception of reality that has no further need to actually see reality as it is "out there" in innumerable variations. Seeing is thus reduced to a monolithic function; it serves to gain global knowledge in a linear fashion that mediates any valid experience of reality into theoretical models, via language.

It is against this background that we must read Serres's proposal for a new way of exercising our sense of seeing: by visiting. Sharing the same Latin root, "to visit" is a frequentative of "to see." This aspect opens a discussion on the local (*Paysage*, 259–79) and the global (*Dépaysement*, 280–83) ways of experiencing reality. Visiting allows us to dwell on and experience details without regard to an ordered arrangement or logical sequence. Serres favors spatial images, plots of land, gardens, landscapes, pages, all related to the Latin *pagus* (which, not so incidentally, he also connects to "peasant" and "pagan") to teach us how to see by visiting, patiently, one locality after another. Each implies of course a certain attention to what lies haphazardly and ill-defined beyond its borders. The local is not effectively visited unless it is seen in its circum-stances.

This is the Serresean implication of visiting: seeing circumstantially with all the sensate parts of our body; in other words, patiently circumnavigating every locality encountered, and in this manner sewing together the strewn circumstances of reality (261). On the other hand, "'seeing' without 'visiting,' seeing only with the eye, the proud sight of pure theory and its inevitable complicity with language, brings with it the effacement of a reality which is always local, multiple, circumstantial, and *mélangé*" (Kavanagh, 939).

Visiting requires patience and above all time, not the record time of shortest distances between two points, measured in terms of road markers and architectural building blocks, but the time one spends wandering along the zigzag path of grazing goats on a mountain slope, an ambling and twisting route laid out around random obstacles, rocks, cracks, hillocks, protrusions, cairns, and mounds (265). The first is a linear seeing that plows through the countryside, marking it as with a straight road. This Cartesian method efficiently subsumes all local details to arrive at a global knowledge of reality, in the shortest time and by the most linear route. In its wake, it erases all local plots and levels all singularities and deviations. It erodes the landscape of empirical knowledge (reality) in order to achieve, as efficiently as possible, a theoretical model of reality that we then take to be reality itself (280).

By using the term *dépaysement*, which Serres wants us to read literally as "de-landscaping," he links the Cartesian method and its eroding effects to a fundamental loss which we should describe as ontological, because in contemporary French *dépaysement* refers to a feeling of uprootedness or being at a loss. Pure Cartesian rationalism uproots the human subject and separates him from the space of empirical reality his senses can no longer recuperate or revisit. We see intellectually, and form ideas geometrically. This seeing produces a method by which we gain a theoretical knowledge of the world. The zigzag wanderings, on the other hand, immerse us in a sort of hands-on experience with space. No distraction is too insignificant, each makes us leave the straight road and explore the byways of a new landscape or another garden, the rock behind it, the plot around the corner. Our voyage becomes a *randonnée* (etymologically linked to the English random) of the unexpected, the forever variable. The bouquet of a good wine composed of a chaotic variability of empirical data, becomes now a nonlinear, unpredictable process of learning in a landscape of sensate experiences (271).

The notion of visiting as the natural meaning of sight is presented over and over by images of place and space. The place (*lieu*) becomes the generic term for the numerous points in space that Serres visits for us. On the surface of his text-landscape, time seems forgotten in favor of space, the topological space of the *randonnée*. When language recedes, we see/visit the earth, and our haphazard visiting mirrors her fractal nature. At every turn and every locality, we see something new and unexpected which leads to further twists and turns in an unending process of discovery. This is certainly an inefficient method for the "monoculture" of the rational mind, but a richly yielding one for the new knowledge or the philosophy of circumstances. The circumstantial connections between places guarantee a meandering movement; the place is not merely static but implies also a dynamic nature, for it cannot be experienced unless one's visiting also pays attention to its circum-stances. In order to fully understand Serres's discourse at this point, we have to recall what he writes earlier: "I am but speaking of time" (172). In the *conflation of place and movement* through the sense of sight/visit, Serres binds time to space, an alignment we have commented on before but which now takes on specific forms that influence his writing beyond *Les cinq sens*.

The specific discourse on dynamical space is evoked by Serres's regret that Western epistēmē is dominated by linear ordered thinking, and by his vision of a "philosophy-to-come of mixture (*mélange*) [that] unites the global and the local, irenically, and supposes a different ontology" (283). The section "Méthode et randonnée (Global et local)" (284–308) lays out a grid for this "new" ontology. On the surface, Serres's discourse remains in the realms of philosophy and mythology, when he contrasts Cartesian order to "Ulyssean wanderings" renamed "exodic order" (286–87), in the sense of exodus as a road that deviates from the common course. In effect, while following Ulysses' torturous meandering, Serres traces a model borrowed from what he knows is causing ontological upheavals in the scientific realm: nonlinear dynamical systems theory.

His long epistemological search has failed to detect significant ontological changes in the classical sciences or the humanities that would offer the conditions for a reunion of our sensate body with our subjective intellect. Neither does he succeed in finding the necessary conditions for a concomitant reconciliation between the local and the global. Therefore, Serres reaches out to this (relatively) new mathematical field that found its first adherents in physics and related applied sciences, but to this day is not universally accepted and has its skeptics and detractors. The "Ulyssean exodus" is explained and supposed with and through the primary concepts of nonlinear dynamics (287–308). Of course, from its inception the notion of circumstance is a key notion—by other words—of this theory which echoes throughout the text on the senses. But nowhere more than in this passage is dynamical systems theory as often implied and becomes a dynamical discourse that informs the remainder of *Les cinq sens*.

This is Serres's version of the Homeric Odyssey: The "Ulyssean order" is characterized by *fluctuations* and *ordered plateaus of stability* that lie *outside* the normal order; they are called:

> *Strange attractors* [emphasis added]. As if types of order existed that are unpredictable in relation to normal law, and types of equilibria unpredictable by known laws of order. As if all haphazard fluctuations, unexpected storms or stochastically extended climatic instabilities would suddenly lead to the formation of a temporarily stable locality, to an island where another weather

> front forms, locally, forgetful of the original one, of the ordinary course, forgetful of the normal climatic curve. Diverging when compared to the "methodic order," these islands create order by fluctuations, a different order one could call exodic. You never find them by the methodic process; exodic, exotic, and ergodic, they lie (*gisent*) outside the global equilibria of our epistēmē.[19] [Note added] Method minimizes frictions or obstacles and cancels them; the "exodic order" plunges willingly into their disorder. (287)

This passage is not so much a retelling of Homer's poem as it is a scientific discourse that breaks with our accepted epistemology. It presents an epistemo-critical viewpoint that breaks with a long tradition effectively established today in the sacrosanct domains of science and the humanities.[20] It is identical to the laic perspective from which Serres sets out to visit "circumstances" rather than insisting on invariant *stasis*. The scientific part of his discourse is an unspoken confirmation of his debt to nonlinear dynamical systems theory. It offers him a way to present the new knowledge while at the same time demonstrating that the new is, in fact, an ancient "savoir" passed on through the epic beauty of Homer's poem, a global landscape of unexpected localities.[21]

Today, we enter a new phase which allows us to combine the local and the global. Thomas Weissert, for example, suggests a realignment of nonlinear dynamical systems theory to allow a view of our "cultural matrix, [where] we begin to recognize similar patterns of dynamical structures" (Weissert 1995, 112). Like Serres, Weissert seeks to reconnect the scientific theory to the broader panorama of human experience. He underscores the importance of the paradoxical union of the global and the local that characterizes the behavior of nonlinear dynamical systems, while in the past "[t]he possibility of a harmonious combination of both pattern and randomness together was not anticipated in mainstream scientific culture" (Weissert 1995, 124). One of the key notions of chaos theory (by which nonlinear dynamical systems theory is popularly referred to), the strange attractor that Serres finds in the Ulyssean exodus becomes Weissert's paradigmatic "figure for a new epistēmē." These notions allow us to summarize Serres's presentation of the philosophy of circumstances or Ulyssean "exodic wanderings" as an exploration modeled on nonlinear dynamics.

Always exodic in his own discursive strategy, Serres does not directly use the term "nonlinear dynamics," but speaks instead of an activity which unites (Ulyssean) praxis and (Cartesian) abstraction. He continues comparing circumstantial wanderings to trajectories passing through all the points in the space of a given landscape where any particular curve can be defined as a local cross section (296–97). This description portrays the geometric outlines of a "phase space" of a system in its various phases or stages during the wandering or "visiting." Then, while the reader recalls that we are still in the midst of a reading of the *Odyssey*, Serres suggests that

> we draw an interesting path that abandons its optimal course of least resistance (*le talweg*) and begins to explore a place (*lieu*): [this path] does not decide in any predictable fashion, but searches; it seems to ramble, not on its own accord or sure of itself, but restlessly, far from its equilibrium and without respite. It is on a quest, always on the lookout, covering the territory, probing, prospecting, reconnoitering; it scours the countryside, hopping and skipping here and there. Few things in this space escape its sweeping probe; whoever follows or invents such a path risks losing everything or inventing something; if he does make a discovery, his path will be known as one which left the straight course in favor of strange attractors. (297)

This passage describes the phase portrait of a nonlinear dynamical system, in *nonscientific* language. Transposed into mathematical terminology, it says in fact:

> The key to the geometric theory of dynamical systems . . . is the *phase portrait* of a dynamical system. The first step in drawing this portrait is the creation of a geometric model for the set of all possible states of the system. This is called the *state space* [or phase space]. On this geometric model, the dynamics determine a cellular structure of basins. . . . Within each cell or basin is a nucleus called the *attractor*. . . . The history of a real system [is] represented graphically, as a *trajectory* in [the] geometric *state space*. (Abraham, *Dynamics*, 13, 21)

Such trajectories settle down to an observable behavior called "their limit sets" which are the attractors (Abraham, 42). The settling down is a geometric way of modeling the evolution (or timing) of the system so that one can predict its future course. In fact, Abraham speaks dramatically of "prediction forever" (29). No less emphatic, Weissert explains:

> [The attractor] is a unique convergence of many and often all trajectories in a system of trajectories, such that we say that the attractor is the ultimate fate of the system. . . . If it were merely the case that dynamicists think of the future as that which draws the present towards it, then there would be nothing significant about the attractor. We make the distinction between any future state of the system and the attractor because the latter has a unique identity transcending that of any single trajectory. An attractor obtains its identity from the union of all initial conditions [of the system] whose trajectories arrive there; that union is called the "basin of attraction" for that attractor. (1995, 118)

Depending on the kind of attractor, either the prediction for the future behavior of the system can be made with great precision (simulated by point or static attractors, and limit or periodic attractors), or predictability is lost when a system becomes chaotic. When such a chaotic system is simulated in its state space, any one of its trajectories falls onto a *strange attractor* which "guarantees a set of global limits on the range of motion of the system, but within those limits, we cannot know exactly what the trajectory will do locally" (Weissert 1995, 122). Since the future of this trajectory is apparently chaotic, Abraham speaks of a "chaotic attractor" (*Dynamics*, 294) and describes unpredictability in this context as the "total ignorance of the position of the trajectory within the chaotic attractor" (303).

It is within these definitions pertaining to nonlinear dynamical systems theory that Serres anchors his description of the Ulyssean wanderings or the philosophy of circumstances. The discourse on the five senses may be read as the various dimensions that Serres needs to set up a sort of discursive state space for the circum-stance, perfect expression for the Ulyssean "exodic order" and perfect linguistic double that supposes "bounded randomness [indicating] both determination and randomness" (Weissert, 122). The philosophy of circumstances is the love of wisdom that rejoices at the variability, unpredictability, and uncertainty, implied in the disquieting presence of strange attractors (*Les cinq sens*, 297).

Serres then guides the reader through representative domains of our epistēmē (classical logic, classical grammar, the natural sciences, and psychology) and shows that circumstances are the real but forgotten condition in the formulation of their principles (308–37). In these examples which make for exhilarating reading, the circumstantial makes possible an exchange and union of global universality and local singularity (333). *Circumstance* describes at one and the same time the difference and sameness of seeing and visiting. The paradox implied in this kind of seeing is lifted by the (no longer) oxymoronic circumstance. The traditional Hegelian philosopher sees the tree in bloom; his seeing is static, is -stance, like a statue. But in real life, we seldom watch so motionless, we find ourselves and the world in constant movement. We pass from one locality to the next. Even at the core of statics and rigorous thought, Serres thinks circumstances when he concludes, "I think if and only if I carry my disquieting doubts into regions where the weighing-thinking (*la pesée*) runs risks" (316). Visiting is dynamic, is circumstance, is random roaming through the empirical countryside. The mixed body (*corps mêlé*) roams the earth visited as a mixed space (337), as an inflated locality (338) or as universal variability (339). In contrast to the motley social events that Serres describes in *Rome: The Book of Foundations* and which lead to a historical transcendental place forever tied to the static concept of *pax romana*, *Les cinq sens* proposes a transcendental *geography*. It is yet another apologue that spatially tells of a new time.

The epilogue called "Joie" describes a life steeped in the mixture of sensate experience. Obliquely personal yet curiously universal, the local and the global bound together by style, it relates experiences of a childhood that make for more than a conventional *joie de vivre*, rather an angelic state of ecstasy. Taking a liberating breath, leaping as if one's feet were forever bouncing off a trampoline, running until it feels like flying, flying and knowing how to make a lover fly, all these experiences unite into a sensation of pure, empirical joy. Serres speaks of an angelic state in which our sense of ourselves as subjects is now joined with our sensate bodies. The knowing subject and its newly knowing sensate body now complement each other, where—ever since Plato separated them—they excluded or ignored each other.

> The knowing subject finally occupies his house, truly and completely, all of his old black box that some ancient thoughtless cruelty reduced to an absent hole, from which he had been excluded and chased, without hearth or home, exiled from his body and alienated from himself. Why was he pressured to renounce with great spite the ancestral home and destroy it through reasoning and scientific processes? (360)

When the subject finally reunites with the body, it is no longer a mere house of language, but a sumptuous one of complete and whole knowledge, "delivered from rivalries, a body-subject, consenting, illumined, transparent, vibrant, spirited, supple, agile, alive: thinking (*pensant*)" (361).

The strange dilemma of the black box is resolved, not by methodic analytical explanation, but by a circumstantial, exodic order relying on the unexpected, the variable, the mixed and haphazard in local experiences. To understand the black box is to exit from the linear path and follow the inventive movements of a zigzag course into the chaotic uncertainties of the world. The inefficiency of the nonlinear path runs counter, of course, to the economy of time dictated by the age of science. But trying to save time is like trying to stockpile what cannot be accumulated, retained, or integrated. The time saved up "in banks" is never the one deposited there in the first place. The discussion of the five senses demonstrates that we cannot seize time as a sum total or even as a series of subtotals. It invites the reader to roam the topological space of localities where one's fluctuating wanderings are the new expression of time (360–69).

Serres implies a topological equivalence between space and time, through the "logic of circumstances" derived from a renewed understanding and praxis of the senses. The preceding discussion of nonlinear dynamics was meant to facilitate this crucial understanding that grounds local randomness and global determination into a field of qualitative wholeness, in the absence of quantifiable predictability. Serres himself focused on the state space of nonlinear dynamical systems as a particular topic of complex dynamics. It is by a *space* portrait, of the evolution *over time* of a system, that dynamical systems theory has arrived at insights that will impact on future epistemological strategies. Since *Hermès. La Communication*,

Serres has been guiding his discourse to this threshold, following a Ulyssean course over multiple texts. The urgency of "supposing" a renewed notion of time seems to drive these writings. Increasingly, Serres expresses his conviction that the rationalization of time leading to its historical conceptualization is at the heart of the postmodern epistemological crisis we are condemned to live today. The merit of his reflections on the senses lies in the production of the simplest but most effective implication of ahistorical time:

> The immediately given (*le donné*) passes quickly by; it is free because it is instantaneous. Sensate pleasure lasts but a moment, a tiny differential of time. In the fleeting sensate lies the infinitesimal disappearing of time. The aesthetic makes the moment shine forth or flare up; it assembles the set of inchoative instants. (363)

For Serres, the chaotic mixture of sensate experiences is intimately wedded to the infinitesimal differentials of time. Only in this union are we capable of knowing both reality and time. Those who digress and embark on an exodic path visit the innumerable localities of space; by thus losing or wasting time they take their time to find real time as a non-integrable aggregate of time differentials (363). In *Les cinq sens*, time is reduced to its essence: the multiple, variable, and unpredictable moments of infinitesimal duration that we experience through our senses. Conversely, through a nonlinear dynamical model Serres shows that time cannot be grasped globally as a calculated sum total of localities. Time summed up in this manner always becomes history or linear science.

History (or the humanities) and science replaced the world of mythology and religion, but they merely replaced one form of the sacred with another. By patiently implying a new way of experiencing time, Serres delivers it from the stranglehold of the sacred expressed in our era by rationalism and technology. This is what he means when he characterizes his discourse as the expression of a laic viewpoint. It effectively frees him from the axiom of science as a pure domain. Science deems its objectivity pure if and only if everything else is excluded from its "sacred" ground, thus renewing the cycle of contention, jealousy, and cantankerous clamor. Just as he implies the black box of the senses, Serres supposes science in an absolute sense, by which he means putting science in a subject-position: Science as an object worked outside us, but now science is immersed in us, we digest science, it becomes one in and with us. Therefore it is no longer a sacred, exclusive domain; it has become part of our laicism (371). The text on the five senses does not pretend to discourse *on* science. But within its discourse, science works humbly and diligently to reconcile subjective intellect and objective reality, without insisting on its position of privilege.

The old division of inclusion/exclusion or sacred/profane is no longer valid. This new situation offers conditions for a philosophical laicism abandoned long ago when words and signs took over the place of the subject and we became creatures of language. The digesting of language as a phenomenon took the time of our entire history. Something similar is now happening in the realm of science, only at record speed, almost abruptly. Serres no longer understands the space of science as a sphere of action for the initiated few, but as something borne in us both individually and collectively (372). The scientific era will have truly begun only when we have forgotten that we live it, in us as subjects, and outside in our sensate experiences of the objective world. The task is to build the sphere of science as a house of laicism where no shaman draws lines on the ground to mark off a sacred space. Serres himself builds its model in *Statues* and implies its "new" inhabitants in *Le Tiers-Instruit.*

In order to accede to the laic state, we have to free it from the two hegemonies of language and science by internalizing them. Serres calls it a "digestion," and a process of forgetting (373). Through its absorption into our subjective being, we become one with it, in the way we have long ago become identical with language. Serres's dilemma is brought about by his attempt to use language as a means to break down the prison walls of language, and to desacralize the space of science as a pure space of privileged subjective activity: "This I have labored for with passion: that in my books knowledge and science would plunge into forgetfulness, so that their very loss may work towards new objects and bear a new subject" (374). *Les cinq sens* celebrates this forgetting by placing its discourse in the realm of aesthetics where philosophy began. It promises the renewal of philosophy for a new subject who, armed with all the internalized knowledge and science of a long and fruitful history, is *free to invent.* Serres does not know where invention will lead. But to the new thinking subject he can offer the very first true object of the new knowledge to be invented: the immediately given. Detached from, and forgetful of the burden of his old thinking, the new subject soars into an empirical reality which is unforgettable (381). *Les cinq sens* is the springboard for such a leap.

4

Time Dynamical: Reading *Statues*[1]

> I tell you: you must have chaos inside to give birth to a dancing star. I tell you: you still have chaos in you.
>
> —Nietzsche, *Thus spoke Zarathustra* I

In order to embark on a reading of *Statues* that respects the evolving wholeness of Serres's writings, we have to summarize briefly some of the implications he arrives at in *Les cinq sens*. Renewing our sensate powers allows us to understand time as infinitesimal "differentials" which together cannot be totaled up; they are non-integrable. Always inchoative, they are local events we grasp or "suppose" through our senses. When Serres writes about the sense of smell, he is particularly emphatic in showing the multiple variability of a fleeting odor, an amalgamate of infinitely many conditions, so unstable and chaotic that it never repeats itself exactly. At any given moment, it is a singular differential of all the spatial and temporal conditions that produce the particular odor or taste. All possible differentials remain inchoative, but collectively we call them "memory," which is a sense of time attained through the senses. Serres's use of the term "inchoative" is significant, for it allows us to recognize moments or time segments which are independently and actively original, each a *pagus* or place if we choose to model this notion spatially as a topological *lieu*, as does the author. Time-as-history, however, is produced when these temporal localities are integrated, by a process which applies a conceptual framework analogous to that of analytical integration in mathematical calculus. In contrast, *the patient visiting of one place after another produces a "non-integrable time"; it is an "empirical" time identified by sensate movement through space, as inchoative instants*. By modeling this timing of space with the help of the Homeric epic, Serres proposes a philosophy of circumstances.

The notion of circumstance functions for the reader as a bridge between *Les cinq sens* and *Statues*. Circumstance, or the "circumstantial" in the adjectival mode favored by Serres, is the paradigm for a Ulyssean order. Set in opposition to the Cartesian method, it is a nonlinear, unpredictable process of evolution toward some meaning. We may call it "islands of signification," since Serres is using Ulysses' meandering sea voyage as a privileged example. In the foregoing chapter, I have repeatedly underscored the double significance of circum-stance combining the stable or *stasis* with variegated, mixed, and always unexpected fluctuations which surround or *circum*-scribe ordered stability. But this is not a "normal" stability; it is an "exodic order" merging with, and emerging from the disorder of circumstances.

Evolving from the parasitic paradigm of earlier Serresean works, the theme of the multiple or *mélange* has been amply elaborated. However, the approach to stasis which is now emerging "out of the ordinary" reveals a new facet and signals a further stage in Serres's discourse. We must recall that in his writings stasis has been the logical opposite of mobility and fluidity, of disorder, of exchange, distribution, interference, passages and communication. Since *Hermès*, Serres reminds his reader that the statue derives its name from the Latin *stare* (to stand) and is the perfect paradigm for stasis, order and rational action, but equally so for immobility, stagnancy, and rational tyranny. It is the incarnation of all that is static.

The statue as emblem of stagnant inflexibility undergoes a paradigmatic change in the logical aftermath of the shift from a "static" order of the *stans* or *thetic* to an "exodic" order of the circumstance. The two "hemistichs" of the term "circum-stance" both separate the static from the dynamic, or order from chaos, and unite them into one concept. The two halves are reiterated in more scientific terms and described together as a strange attractor (*Les cinq sens*, 287). I have alluded to this strange new statue in chapter 1 of the present study. Now I propose to call it a "dynamical statue," oxymoronic for Cartesian thinking but meaningfully "supposed" within the logic of circumstances. The statue as circumstance is no longer either purely stable or utterly chaotic; it is both circum- and -stance. In terms of dynamical systems theory Serres himself has recourse to when discussing the circumstance in *Les cinq sens*, the Serresean statue becomes dynamical and may be supposed to model a strange attractor.

Willful as this supposition may seem, it becomes less so if we understand that the properties of strange attractors are unknown and, in fact, remain *unknowable* on the local level. For classical mechanics such attractors do not exist. Systems displaying such propensities are considered as drifting toward disorder and therefore not suitable for scientific analysis, unless they are linearized, that is, adjusted to fit predictable outcomes. But for nonlinear dynamics, computer simulation of nonlinear systems in evolution has rendered the unknowable a fascinating object of study, for it becomes "visible" on the monitor screen: Trajectories that graphically indicate the behavior of such a system can, when it becomes chaotic, merely circumscribe its individual local stages or phases. At the chaotic stage, they no longer identify the local phase with quantifiable precision, but trace out a region called an "attractor basin." Globally determined by this area of probable yet locally unknowable behavior, an attractor is traced, only traced and nothing more. When first observed, this attractor was quickly dubbed strange, because it is traceable but unknowable, globally determinable but locally unidentifiable or unnameable. In other words, it offers determination and uncertainty at one and the same time.

The concepts of nonlinear dynamics reflect not only new and different mathematical perspectives on reality, but also reveal the necessity to redirect traditional philosophical conceptualization. When pleading for a new and different way of thinking, Benoit Mandelbrot and other dynamicists forcefully imply a much larger conceptual field than their purely mathematical or scientific domains. From this perspective, nonlinear dynamical systems theory is a scientific theory that makes extremely efficient metaphorical use of embedded philosophical and ontological concepts (being, presence and absence, identity, time, freedom, determinism). The popularity of chaos theory among nonscientists reflects the success of this modern "metaphorical language" (we normally refer to as mathematics and science) to persuasively translate age-old philosophical questions into postmodern thinking. Serres's texts under discussion in the present study reflect this complex and interrelated discourse. However, the specificity of Serres's thinking lies in his ability to move back and forth between various "metaphorical languages," while keeping in focus the fundamental philosophical issues involved. Since the notion of time is being traced in Serres's texts and time is central to dynamical systems theory, I try to establish a close conceptual relationship between the two.

The use of scientific theory in general, and chaos theory in particular for studies in philosophy, literature, and the social sciences has been and is a controversial issue focused on the extent of transfer of terminology, concepts, and models from one field to another. H. W. Brands, for example, cautions against too hasty or simplistic a matching of notions of chaos theory and studies in history, yet does not reject in principle such readings for historicist analyses. "[The] notion of fractals . . . might provide historians something to ponder, although their usefulness will probably be greater as metaphors than as models" (Brands, 500). The value of notions such as fractals and what they entail for historical studies lies more in the recognition of multiple approaches that are equally valid (Brands, 501). For the historian, the difference between science which gives descriptions and history which offers explanations remains problematic (Brands, 509). Patrick Brady is just as cautious about the applicability of nonlinear dynamical systems theory to studies in the humanities. In a very balanced article in which he recognizes, nonetheless, potentially useful overlapping domains of interest, he concludes: "Chaos theory . . . raises the question of extrapolation from one field to another, and the status of such extrapolation. It is easiest when reduced to fairly general principles, but more insightful when based on an imaginative use of analogies" (Brady, 77).

The question whether the present study uses nonlinear dynamical concepts in strictly scientific notation, as models, or as metaphors is an important one. It also touches on a more specific question about the proper approach to Serres's discourse and comes now to the forefront, since his text *Statues* is the subject under discussion. The connection between the scientific and the epistemological theories will develop as both an analogous and metaphorical one, with the added reminder that in the Serresean context metaphoricity carries the same cognitive value as other approaches. In the sense that I understand Serres's dynamical discourse on the static, as it develops in *Statues* and *Le Tiers-Instruit*, and the scientific description of the world via nonlinear models as two metaphorical discourses on the same underlying philosophical concerns, I do equate the two.

The conceptual foundation of *Statues* is the discovery of dynamical inventiveness at the very core of stasis and is thus significant for dealing with the problem of evil (in the form of death) and the "new" time. The many functions of the statue are presented in the narrative via a series of descriptive portraits of historical/cultural phenomena that trace out the dynamical behavior. *Statues* is the discursive monitor screen that renders "visible" the chaotic as a fascinating object of study for nonlinear dynamical thinking. What the text presents, then, and what I invite the reader to consider, is the crucial intersection where Serres's quest for a true understanding of time has arrived at a point which, seemingly farthest away from scientific discourse, parallels the most recent scientific and mathematical findings concerning nonlinear dynamical systems.

Reading *Statues* is a process of discovering the passages and implications of Serres's "non-epistemological, scientific discourse." It is of no small consequence that sculpture as a domain of artistic or aesthetic activity is here implied, coming on the heels of a study on the senses as the original object of aesthetics. With its model of non-Cartesian order and non-epistemological discourse closely related to concepts of dynamical systems theory, the discursive strategy of *Les cinq sens* continues into *Statues*. While visiting some of the statues discussed in this text, I hope to lay bare the philosophical and ontological implications that are the same conceptual basis for the mathematical "strange attractor" as well as for the figure of the Statue in *Statues*. In this sense, and this sense alone, the dynamical statue simulates the strange attractor which, as I try to show, inhabits *Statues*. The object of this chapter is therefore an elucidation of both the subject of Serres's discourse and this discourse itself. In the final analysis, the two aspects collapse into one, a discursive portrait of what I have coined a dynamical statue.

The transposition of concepts that dynamicists are still elaborating and refining, to a discourse that aims at our epistēmē as a whole, is perhaps risky.[2] But Serres does not leave much choice, if his reader is to digest the findings of *Les cinq sens*, on the one hand, and lay the groundwork for a reading of *Le Tiers-Instruit* (four years after *Statues*), on the other. This approach also shares more global descriptions and analyses of common aspects shared by postmodern texts and dynamical systems theory:

> I make the analogy between a text—considered as a complex trajectory space of narrative and signification, generated by an author's conscious and unconscious model of cultural relations—and the phase space of a dynamical system, generated from the dynamicist's model of the relations among the degrees of freedom of the referent physical system. (Weissert 1995, 124)

Although Weissert's analogy limits its cultural component to texts in which deconstruction is at work and Serres, for his part, shies away from postmodern literary theory, *Statues* contains elements of an epistemo-critical "deconstructive" reading of the postmodern epistemological scene. This text is best understood if it is treated as a discursive phase space that simulates the evolution of Serres's "non-epistemological, scientific" look at modern culture.

There are also more practical considerations for applying complex systems theory to a reflection on Serres's text itself. They go beyond the author's own use of such concepts in *Les cinq sens* where he describes the relationship of our subjective intellect with empirical reality. First, such a reading strategy offers an alternative to the sometimes perplexed impatience that seems to creep into reviews of Serres's works that are under consideration here. Perhaps the best example is Thomas Ferenczi's discussion of *Les cinq sens* (*Le Monde*, November 15, 1985) in which the critic urges the author to bring his chain of metaphors to a closure and "hasten his step a little" in view of some definitive answers (Ferenczi, 30). Ferenczi's remarks indicate how well he has read Serres, but how little he has understood him. For he welcomes his tireless efforts to indicate "the directions to be explored," but deplores the lack of a definitive conclusion as to what lies at the end of that road. The critic has grasped the pattern of Serresean thought but regrets that there is no stable point solution. He has grasped the emergent properties of this discourse but is not satisfied with such an open-ended system, preferring instead the linear attractor of a final, definitive word. An in-depth reading of Serres's texts at issue here must take into account, however, their nonlinear structure.

Second, in order to recognize the relevance of dynamical systems theory to a global understanding of *Statues*, the reader should be aware of the similarity between the paradigm shift indicated by the "circumstance" and the one revealed by the new mathematical-scientific theory. Once again, Weissert gives a succinct description of what is involved here:

> In order to recognize the relevance of dynamical systems theory to the project of interdisciplinary studies, we must differentiate between "dynamic structures," which are those undergoing change but whose structure may be said to exist in stasis [and which classical mechanics deals with], and the much more abstract "dynamical structures," which are *structures of change or movement* and *which cannot be given form without reference to time* [emphasis added]. (Weissert 1995, 112)

This echoes the introductory chapter of *Order out of Chaos*, where the "new and unexpected role" that time plays in postmodern science is stressed again and again, together with self-organizing systems and complexity (Prigogine and Stengers, 10). For these authors, time is no longer reduced to a parameter in the static view of nature as espoused by classical dynamics. Rather, time is now understood as a pivot in the field of the new dynamics which "may be defined as the study of the movement of signification through time" (Weissert 1995, 114). Concepts of dynamical systems theory underpin Serres's concept of ahistorical time, whether the latter is presented through apologues as in *Detachment* or empirically aesthetic as in *Les cinq sens*. Equally important, the issue of fuzziness in the texts under consideration here, and critiqued by the reviewer of *Les cinq sens* in particular, is not a tangential issue of style or, worse, confused perception. Rather, it is fundamental and critical to the communicative process and transmission of knowledge, as Ferenczi rendered so explicit in his review, albeit from a negative viewpoint.

The first of the new breed of dynamicists had to defend themselves against similar criticism for relying on computers, simulation, and numerical integration for their experiments, the criticism coming from more traditional mathematicians who would ask proofs and demonstrations, sometimes with a good dose of malice. Furthermore, dynamicists were (and are still today) exceedingly conscious of the critical importance of their unconventional methods for a formulation of fuzziness or the "unknowable" itself. James Gleick, for example, writes in regards to Hans Peitgen, mathematician and specialist on Mandelbrot sets: "Granted, . . . [t]o see an image on a graphics screen does not guarantee its existence in the language of theorem and proof. But the very availability of that image was enough to change the evolution of mathematics" (*Chaos*, 230). Quoting Peitgen himself, he continues: "Rigor is the strength of mathematics. . . . But you can look at situations that can be understood *partially* now and with rigor perhaps in future generations. Rigor, yes, but not to the extent that I drop something just because I can't do it *now*" [emphasis is Gleick's] (*Chaos*, 231).

In *Genesis*, the various metaphors Serres uses to express chaos introduce a qualitative alterity that dissociates chaos from what is basically its binary connotation with disorder. Disorder can only denote a quantitative (negative) difference to order. Serres coins the term "positive chaos" and implies its definition by pursuing a long metaphorical chain that now produces a dynamical statue. Absorbed and enchanted by this poetic and very figurative language, the reader should not ignore, however, the many, often allusive references to various elements of complex dynamical systems theory, which are more explicitly used in the later passages of the text on the senses. Serres explores the chaotic implications of Ulysses' travels, hoping to present a zigzag logic suitable as a *pharmakon* against the Cartesian method which impoverishes our epistēmē with its insistence on unitary purity. In *Statues*, he models his epistemo-critical discourse into the "landscapes" or the *pagus* of dynamical systems simulations: In the wake of *Les cinq sens* and its story of strange attractors, *Statues* becomes a textual phase-space, that is, a discursive portrait of the phases of our epistēmē as a dynamical system.

Phase-space (or state-space) is a virtual space that must not be confused with the "usual physical space," as Monique Dubois and Pierre Atten carefully underscore in an article on "chaotic order:" It is an abstract, mathematical space whose multiple dimensions correspond to the number of variables (or "degrees of freedom") that compose the state of a given dynamical system (Dubois et al., 193). However, in the transposition of concepts I am proposing, Serres's phase space is a text, and the portrait is discursive rather than a graphical image. Since *Hermès. La Communication*, the statue has always functioned for Serres as the paradigm for a systemic state. "Immobility, inaction, fixity, thesis," these expressions signal the sine qua non of any system based on binary propositions and the principle of identity (*Les cinq sens*, 315). Condemning rationalism for its tyrannical grip on thought and intellectual creativity, Serres perceives the statue in *Detachment* as a veritable fetish symptomatic of the incessantly growing desire of reason to master the empirical world and transform it into a conglomerate of objects of contention. In earlier texts, Hermes and the parasite function, each in his own way, to check or challenge the menacing power of order and stasis.

But after his meandering visits (*la randonnée*) through sensate experiences, Serres makes a mental leap in *Statues*, which allows him to examine the statue, not in its function as a purely static entity, but as a dynamical system. We understand the latter to "consist of two parts: the notions of a state (the essential information about a system) and a dynamic (a rule that describes how the state evolves with time)" (Crutchfield et al., 49).[3] In this conceptual frame, the statue looms before the reader as if it were "the dynamical behavior at the core of any time-dependent phenomenon" (Dubois et al., 193). With this epistemological upheaval of the conceptual statue, Serres achieves three things: (1) He resolves an ambiguity, concerning stasis, that lingers in some of his earlier allegorical figures (for example, Hermes or Diogenes' snowman); (2) interpreting the roles various "statues" play throughout history, Serres creates an admirably coherent scene of our cultural and epistemological evolution; it is a story that mixes History with a "history of things" (*Conversations*, 165), at times producing scenarios that shock because of their stark or almost brutal juxtaposition of the most disparate events; and (3) in direct opposition to unitary, tyrannical, stagnant, and mortal rationalism, and in contrast to what he had always described as "statues" in the sense of absolute stasis or immobility, Serres now constructs a statue that, for the reader, acts like a dynamical system. In *Genesis*, he promised to present a text on time; in *Statues* he has succeeded, by way of a dynamical model—in the most unexpected place (*lieu*): at the heart of the statue.

Serres follows a given historical event (the tragedy of the *Challenger* space shuttle, mummification as practiced in Pharaonic Egypt, the petrification of Lot's wife, a car moving along a serpentine mountain road, the writing subject called "Serres," among others) and traces its function as it continually evolves with time, rather than giving a mere thetic (static) description. In so doing, he outlines a process which for the reader takes the shape of a discursive "dynamical" statue, that is, a statue whose function changes over the time of its usefulness for the creation of new concepts. Each of the event-statues described in the text becomes part of this "structure of change or movement" (Weissert 1995, 112). We must keep in mind the distinction between "dynamic" and "dynamical" structures to mark the difference between classic dynamics and nonlinear, complex dynamics (see p. 107). Both share one object of study, the dynamical system for which Crutchfield's definition is exemplary (see p. 109). But there are two types of dynamical systems; predictable systems are studied by classical dynamics, while unpredictable ones are the focal point of nonlinear dynamical systems theory. A brief description of both is necessary here, for both play a role in our reading of *Statues*.

A system is predictable if its behavior is periodic or linear, for example, a simple pendulum in motion. Its state is a mathematically defined combination of two variables, angular position and velocity; it evolves over time. All possible future combinations of its position and velocity are predictable: either the pendulum repeats each of its positions and rates of change of velocity at regular, periodic intervals, or its motion slows continuously until the system comes to rest. One can represent the evolution of such a system by a portrait, a graphics image of its orbit composed of all possible combinations of the two variables for the duration of the system's motion. The trajectory of such an orbit is plotted in the system's phase-space made up of all points that may represent possible states of the system as it evolves over time. Since a simple pendulum in motion is defined by two variables, its phase-space is two-dimensional, each variable plotted along one dimension or axis, and resembles, in form only, the Cartesian coordinate system we are all familiar with from high school mathematics classes. The trajectory tracing the behavior of the pendulum converges either towards a limit or periodic cycle, if the pendular motion is sustained by an exterior force, or towards a fixed point, if the motion is not sustained and dies out at a regular rate of decrease in momentum. The limit cycle and the fixed point are "attractors" to which the trajectories of the system orbits are drawn. The future behavior of such a dynamic system is therefore predictable and contained in, or summed up by, the attractor.[4]

We now transpose the concepts of such a mathematical model of systemic behavior to the reading of *Statues*. The juxtaposition of the Carthaginian cult of Baal, on the one hand, and the American shuttle disaster, on the other, serves to initiate Serres's discussion on modern epistemology in crisis. Transcending all individual events of biological death, human history is the construction of religions, mythologies, traditions, and cultures; together, they form a "collective immortality," in the historical sense of this expression (22).[5] The construct functions on the basis of two interconnected activities in space and time: "Efficient operating techniques construct the world, and socio-cultural technologies construct time" (23).[6] This formulation effectively defines our epistēmē as a dynamical system for which the information of its *state* is given by the "operating techniques" (of pre-Cartesian and modern natural sciences), that is, by a set of determined and interdependent data or variables. On the other hand, the evolution of this state over time (its *dynamic*) is initiated and nurtured by "the socio-cultural technologies" (the human sciences).

But, writes Serres, the crisis of modernity is born of a creeping forgetfulness concerning the dynamics of the construct, because we set the natural, so-called exact sciences in opposition to the humanities, and vice versa. Our era favors "the efficient, operating techniques" which evolve into "industrial techniques," at the expense of the "socio-cultural technologies." For our reading, this means that modernity favors the system's information (i. e., it favors its steady state) at the expense of the system's evolution over time (i. e., at the expense of its dynamic). The construct is no longer recognized as essentially a dynamical system. Therefore, Serres maintains, we no longer know how to contribute meaningfully to the "collective immortality." It is an ominous collective failure: Since Hiroshima "collective death hangs over us, in the global sense of humanity in its entirety" (23).[7] In order to counteract this forgetfulness, Serres attempts to identify the cultural forces of the past, mythology, religion, and art. He does not intend to recover their knowledge of time, the very process of which would produce a historical time, as *Rome: The Book of Foundations* had demonstrated. Instead, he is attempting to restore the dynamics of contemporary stagnating culture. In so doing, he discovers that his task is easy, because history has left behind landmarks that point to, and gather up, subsets of the system "epistēmē." Searching for clues, Serres recognizes them in statues, in the most varied, intuitive and inventive sense of this term as he applies it. We may say that his quest becomes a strategy for modeling the phase-space portrait of a dynamical culture, and that he finds it in the statue. We may further read each statue (the rock of Sisyphus, the space shuttle, stelae, the Black Stone of the Kaaba, the bullfighter, and others) as if it functioned like "an attractor that gathers into itself a given system's internal logic through time" (Greening, 4).

In classical mechanics we have a comparable forgetfulness, but it is known under different names. For celestial mechanics, for example, Newton's laws of gravitational attraction describe completely and adequately the motions of heavenly bodies, expressed in the image of the "clockwork motion of the planets." But Newton's equations can be solved analytically only for the motion that involves two bodies; for the so-called three-body problem these same equations cannot be completely integrated or solved. This means that long-term (very long-term) prediction cannot be made about the motions of three (or more) planetary bodies attracting each other. Computers can increase the scope of numerically calculated orbits, enough to make interplanetary space craft probes possible and realizable today, as well as to predict the course of comets.[8] Bracketing the three-body problem is somewhat similar to forgetting that Newtonian laws applicable to macroscopic systems do not conform to the laws of quantum theory that concern motion on the atomic level but which cannot predict the future for macroscopic systems. This is a kind of "forgetfulness" necessary for practical applications of Newtonian mechanics on which all modern technology is founded. But "the price paid for the deterministic, time-reversible Newtonian paradigm is a maniacally self-deconstructing universe," states Alexander J. Argyros who connects the clockwork view of celestial mechanics with Laplacian deterministic predictability and universal entropy, and labels the Newtonian-Laplacian view as ultimately pessimistic (*A Blessed Rage for Order*, 257). Serres expresses a similar view when he writes about modernity's creeping forgetfulness and consequent risk of collective death. Of course, the forgetfulness of science is wonderfully camouflaged by calculus formalized by Newton and Leibniz, with its limit concept that makes differential equations such a powerful mathematical tool. It is not so much a forgetting as a restriction to solvable or integrable differential equations; in other words, the systems investigated via the tools of calculus must be linear or periodic, they must be completely predictable.

More pessimistically, Serres sees postmodern epistemological forgetfulness as a false perception with "maniacally self-deconstructing" consequences. We may model this with the Serresean immobile statue as a fixed-point attractor for a linear or integrable and, therefore, predictable system. Serres translates this into the attractor "atomic bomb" which draws the trajectories of human history toward its most absolute and rigidly fixed point, global destruction.[9] Mathematical-scientific forgetfulness and epistemological forgetfulness conflate into this fixed point of death. The new science of complexity, with its epistemic shift in conceptualization, offers a (so far mostly theoretical) way out of this cul de sac. Basically, *Statues* addresses the same shift, but couches it in cultural terms, with a view toward connecting modern epistemic activity with past dynamical "industries." In this fashion, Serres theoretically restores the conditions of possibility for the collective immortality he presents in *Detachment* as a fable.

To do justice to a reading of this syndesis via the model of the statue as a strange attractor, it is necessary to speak of the second type of dynamical systems, those with non-periodic behavior. Most natural phenomena are, in fact, nonlinear systems for which mathematical formalism or physical models are difficult to construct, because they tend very rapidly toward chaotic behavior. Their evolution over time is unpredictable. Examples are fluids in turbulent state, or meteorological forecasts for extended periods of time. Periodic, predictable systems are the exception in nature, but science has always privileged them as ideal models for the best and most accurate information about the world. By holding a discussion on the multiple through "apologues," Serres subverts this privilege, thereby risking a certain lack of understanding (both of content and as to style) on the part of his scientific reader. However, his "new" metaphorical strategy parallels the "new" theoretical developments in the scientific community.

The notion of phase-space and its visualization on a computer screen give dynamicists the means to see the orbits of the behavior of non-periodic systems as something more than just random trajectories resembling the dizzy flight of a bumblebee.[10] Plotting the system's complete state at each instant of its behavior over time, such trajectories do nonetheless *not* converge on a precisely determined attractor or solution point; neither do they continue haphazardly in the phase-space. Instead, they trace out an attractor region around which they move in "patterns," now coming closer, now straying further away, but always returning without ever crossing their own paths; that is, the system never exactly repeats a given moment of its behavior and never reaches a determined fixed point or periodic cycle. Any such system, though known in its initial state and determined by perfectly defined differential or difference equations, evolves toward chaotic behavior. Thus, no longer analyzable, only its phase-space portrait can reveal a certain pattern within the chaotic evolution. Inside chaos, then, a certain *global* order or *pattern* is discovered which remains, nonetheless, unpredictable as to its *local* characteristics. Such a system is said to converge or tend toward a strange attractor. The most famous of these attractors is undoubtedly the Lorenz attractor which resembles a butterfly's wings and represents topologically the history of a meteorological system.[11]

Reading *Statues* as a textual portrait or discursive simulation of our epistemo-cultural system's behavior in its various phases, is analogous to the way by which we visually perceive the graphically generated portrait of a strange attractor for a given nonlinear dynamical system. It allows for the discursive expression of our epistēmē as an emergent structure, as theoreticians of complexity call these structures. Their simulation by way of strange attractors is a good model for their inventive richness. This strategy permits us to discover the otherwise invisible patterns of what on the surface appear to be random cultural or historical developments. The suggested approach can be realized only if we understand that each textual statue, that is, each circumstantial system Serres creates around and with each statue-like phenomenon he chooses as examples, implies time on its own local level. In *Les cinq sens*, this mapping is expressed in the notion of "visiting." Reiterated, the text-statues become plotted moments forming the reading orbit within a discursive phase-portrait that simulates the dynamical system of our epistēmē. Moreover, now "time is no longer the global [historical] quantity from which we derive the dynamics. Instead [it is] inserted into [its] formalism. . . . Time is thus an implicit feature [of the system] that we can derive from its dynamics" (Weissert 1993, 29–30). In this fashion, the text *Statues* becomes a subset of its own content-set, because itself is a textual statue aptly named "Statues" displaying chaotic behavior as it develops along its internal logic. As Serres has labored to formulate in his writings since *Hermès*, the dynamical statue modeled in and through this text is the logical consequence of the "transformation of space" (59).[12] Stasis, or the "problem of space" (Harari and Bell), is now reconciled with dynamical time and highlights the epistemologist's overriding concern with modern cultural forgetfulness.

In terms of dynamical systems theory, this neglect has led modernity to the brink of stasis, to a perfectly defined attractor which must be resisted as dangerous and risky behavior on the part of the system: it leads to stagnation, rigidity, and redundancy. For the cultural construct, this means death, as the first two statues of *Statues* so aptly model by their shocking juxtaposition. From the cult of Baal with its murderous rites, to the disaster of the *Challenger* space shuttle endlessly repeated on television screens, humankind has not changed its almost blind dash for death, in spite of technological advances and our passage through the era of Enlightenment. (Lest one object that the *Challenger* disaster was an accident and not a planned course of action with a built-in projection toward death, Serres cites statistics for yearly vehicular deaths on France's highways, arguing that we enlightened moderns regard such numbers, although tragic, as a "normal" and "acceptable" consequence of modern life.) Serres is not content to outline discursive portraits of such statues in view of a convergence toward their ultimate fixed attractor, that is, global destruction as the inescapable and therefore predictable fixed point of our epistēmē. In order to combat epistemological forgetfulness and restore an awareness of our undiminished capability to construct "collective immortality," if only we were to give ourselves a chance, Serres presses on and presents *his* interpretation of a popular science fiction story that anecdotally reveals the close alliance between science and death. The undoing of forgetfulness through the revelation of "constructive" death is a necessary catharsis in the process of regaining the power of collective immortality. The author is moving from two epistemic figures modeling dynamical systems with stable attractors, sacrificial rites of an idol cult and modern space missions, to a dynamical system that visibly shows its instability and simulates a discursive strange attractor. Above all, we must keep in mind that the "statue" we discover in the second *récit* of *Statues*, titled "Second Foundation" (35–52), is also an epistemic figure; however, its simulation as an attractor displays strange and chaotic behavior.

Presenting a unique reading of Jules Verne's *From the Earth to the Moon* (second part, "Around the Moon"), Serres discovers a third element for what in his discourse has always been the two-body system "subject reaching for knowledge of the object (the world)."[13] In the guise of a story about the corpse of a dog, frozen into a statue-like form and flying alongside the rocket on its journey to the moon, the appearance of death turns the predictable system "subject-object" into a three-body system which becomes chaotic. No longer a destructive fixed-end result (e.g. the atomic bomb), or a calculated risk factor embedded in the benefits attained through the most stable components of our epistēmē (mythology and science), *death* joins the system *as part of its determined state*. Three "elements," the specter of the frozen corpse, the subject-scientist, and the object in the form of an exploration of the heavens, become one systemic state in this episode of Jules Verne's story. For Serres, the three components form one epistemic figure, one statue. In our reading, death is one of the three variables that signify this unique figure, death is part of the state (or presence) of this system. As the latter evolves over time, we attain a dynamical system that we call "chaotic," that is, it could display a chaotic or strange attractor. We see this happening discursively in Serres's story.

The canine corpse (symbol of a death innocently told by Verne) holds the complete attention of the subject (the astronaut inside his rocket) who, fascinated by this unexpected companion (Verne's dog is in fact called "Satellite"), abandons the planned observation of the object (the stars). We may say that the deterministic rules at the outset of the voyage do not lead to determination, to the specific and defined data gathering predicted for this space expedition, as the system evolves over the time of its voyage. We are, of course, talking about Serres's Verne-story, which only concerns the episode of the dog, leaving aside the other story with its stable destination, the moon. The destination or fate of the system "dog-subject-object," not necessarily the fate of the voyage or "physical system," becomes unpredictable and chaotic.

Recognizing that the Serresean interpretation of this scene offers a model, not for inchoate disorder, but for deterministic chaos, is the task that remains before we can fully understand the function of the many statues chronicled in the subsequent *récits* of *Statues*. They enter the text as stable systems, but this is only an illusion conjured up by our forgetfulness. Serres's interpretation reveals their unstable or circumstantial core marked by death as an essential instigator or prod (238). As we read the text, death loses its murderous, absolutely static aspect, so that the statue of death (the frozen corpse-satellite) is no longer a static figure; it becomes a model with positive properties for inventive emergent structures that globally form a "collective immortality." This strange configuration is modeled through the strangeness of the dynamical statue.

Death joins the system "subject-object" which the epistemologist knows as "knowledge." Cultural death, and perhaps biological extinction of the species, are ultimate fates of our rationally dominated epistēmē. Binary thinking condemns all cultural activities to an uninventive future with creativity stagnating in monotonous repetitions. This is the result of a faulty interpretation due to our epistemological forgetfulness. Serres's reflections on Verne's scene reveal that the stability of death is sterile only if we consider death as a teleological part of humankind's own activity and as the ultimate product of human rivalrous relations. *Rome: The Book of Foundations* is an account of this role played by death. In *Statues*, Serres refers to it as the first foundation of knowledge and condenses the epistemological story of the foundings of Rome into its nineteenth-century science fiction equivalent:

> Between Barbicane [the astronaut] and his companions, the dog is in excess and must be expelled through the trap-door in the floor, for the sake of the group's survival: [in *Rome: The Book of Foundations*] we witnessed this rite of constitution for the collective subject. (42)[14]

In the history of Rome, Serres traces each resolution of its many crises, each of its sociopolitical renewals, and each of its foundations back to a particular exclusion in the form of murder. Similarly in the Vernean story, the dead animal is expelled through an opening in the floor. The dog becomes in death a "sub-ject," because Serres understands the subject by its etymological components of *sub-* and *-jactus* (thrown) (42).

Unlike the corpses of Rome's foundations, the canine subject-corpse refuses to be buried and forgotten. Serres repeats several times that it acts like a specter returning (the appearance of a ghost is often expressed by *un revenant* in French, literally, "one who is coming back"). Something undefinable accompanies the rocket: it could be a subject (the dog), or death in form of a corpse, or an object (a solidly frozen statue aligned with the rocket), or all three together. Serres explicitly mentions having read Verne's tale in an old illustrated edition with a drawing of this scene. It depicts the astronaut gazing through his porthole at the frozen corpse. But the drawing is executed at an angle that allows the reader-viewer a look at the observing astronaut and at the dog. For Serres, the book becomes the reader's rocket capsule where the written lines are its walls which now and then feature portholes for viewing the outside (38). The drawing is a window allowing the reader to look past the writing (the representation) and to share the astronaut's fascination. As is often the case in Serres's discourse, he deliberately obscures the distinction between the object-discourse, in this instance Jules Verne's story, and his own discourse, which gives us the occasion to conflate both and eventually recognize the strange attractor in both. This will be discussed further in this chapter.

Serres looks at the extraordinary scene and the space it lays open to our view:

> None of the instances in this space has as yet taken up its [assigned] role: the dead dog is object or subject; the observer in his deadly rocket remains enveloped in an object, in the same way that I am in the book and its written lines. Does death stabilize the ties between subject and object? In return, does the object stabilize the ties between subject and death? Individually or collectively, does the subject stabilize the relationship of death with the object, to the point of undecidability?—This threefold stabilization defines the status of statues which are, at one and the same time, concrete and real like objects, and bodies of dead subjects, mummified or haunting specters. (42–43)[15]

These lines make clear that death plays a founding role. As soon as we recognize that it is placed as part of the system at the outset, genealogically, and that it is not teleologically posited, death assumes a role of mediation between subject and object. Consequently, Serres explains that death is not a destructive agent but a generatrix and silent partner of the system "knowledge."[16] The truncated, binary version of the epistemic equation "subject + object = knowledge" leads to epistemological forgetfulness interpreted as the cause of the destructive properties associated with death as an excluded, negative force. But Verne's story models the system with a different equation: (subject ↔ death ↔ object) = knowledge. Paradoxically, death is now understood as a helpmate and instigator for the construction of the "collective immortality" which, in the final analysis, is Serres's label for our epistēmē.

Verne's three-body system "subject-object-death" evolves rapidly toward unpredictability, because the outside phenomenon the spellbound subject studies does not match any of the many local results the subject expects for an ordered accumulation of evolving knowledge (e. g., orderly knowledge of the stars). Usually expelled and silenced, death has reappeared as the third body or element of the system-state (the dog is subject in as much as it was and still is man's companion and best friend, and object in as much as it is now an inanimate and solidly frozen projectile). The system has qualitatively changed from a binary construct to a three-body problem and does no longer produce predictable data.

Serres discerns a perfectly determined combination of three elements in this scene. Yet, he also sees the system evolving into an insight (or knowledge) that is rapidly becoming disorderly. The chaotic dynamical system of knowledge inherent in the Vernean scene is an example of "positive chaos" (or deterministic chaos), described in the following way:

> The objects [the forgotten stars] fade into the background of the real world and the subjects retreat into their box [the rocket], with *an astronomical distance* between them; but the corpse is left *within constant and close reach* [emphasis added] of the walls of the receptacle [the rocket].(45)[17]

The "astronomical distance" stands metonymically for an infinite distance of epistemological recognition, not a physical distance. Therefore, it renders infinite—ergo unpredictable— the evolution-over-time of knowledge, for the system called "exploratory expedition to the moon." In Verne's story the rocket travels on to the moon, but in Serres's reading, the object of the expedition has been sidetracked. Through an encounter with death, another object or petrified projectile, which could also be taken for a subject (the dog once alive and man's best friend), accompanies and mirrors the rocket. The latter is itself an object containing within its hollow a subject and, as we know from the *Challenger* accident, the presence of death. Serres models this complex structure of the "second" foundation of our epistēmē with the help of a statue:

> We do not know if the second foundation precedes or follows, derives or enriches the first [which binds death with the subject]. We know that it binds death with the object. . . . It presents the undecipherable, silent, and invisible face of a founding instance which has no name in any language and which assembles the instances that we separate under the three names of object, death and subject. This fundamental stratum (*gisement*) unites what rests below [sub-], what rests here, and what rests in front [ob-]. Objectifying the subject, death offers the object to the latter on condition that he shape it. What else can one name this stratum, this stable instance, but a statue?[18]

Serres adds that these "statues" precede the appearance of language. Only after human beings acquired language as their abode did they forget and repress the knowledge that the union of subject and object is based on a relationship with death.[19] The seminal signification of statues as funerary signs was forgotten in favor of language and its fictional properties. Circumventing language, statues come back like specters (*revenants*) to haunt or instruct us about the threefold relationship, depending on the degree of our willingness to understand (49). From this perspective, the "statue" of the ill-fated *Challenger* is a direct descendant of the statue of Baal.

Statues is a text that transforms Verne's model into a general epistemological paradigm that, we repeat, includes its own text in its dynamics. In our dynamical reading, we may say that, determined by three elements, subject, object, and death, the story becomes the "seed number" for the discursive system "Statues." One of the fundamental characteristics of nonlinear dynamical systems is their extremely sensitive dependence on initial conditions which can never be known with complete certainty. The unpredictability of such systems' future behavior is a direct effect of this characteristic. We say that their evolution is chaotic. But after many iterations of points (each point representing the complete state of the system at a given moment in time), the simulation in phase-space shows that the trajectories, no matter from what initial condition (or point) they started, will remain within this region of the phase-space. This strange attractor region sets "global limits on the range of motion of the system, but within those limits, we cannot know exactly what the trajector[ies] will do locally" (Weissert 1995, 122). We can now modify our statement about the system's future: It has a *globally* defined future; its evolution is characterized by predictions of "patterns" where local "pinpoint" precision is blurred into "fuzzy borders."

Serres models this "bounded randomness" or deterministic chaos conceptually in the circum-stance, as we have discussed in the preceding chapter. But *Statues* reveals that our collective "technologies" have left other, visible signs that we recognize in the conflated circum-stances Serres calls "statues." Only our blindly rationalizing drive toward linearity has allowed us to ignore the chaotic side of the statue and to focus instead on its paradigmatic stasis or "stable stratum." But in the midst of the modern triumphs of traditional linear thinking, statues stubbornly model the nonlinearity of our evolving epistēmē, because each one, or at least the representative selection Serres gathers and presents in *Statues*, is a global attractor region where subject, object, and death dance a *pas-de-trois* that creates fuzzy but inventive borderlines propelling our knowledge to truly "new," not repetitive ventures.

Discursively, the iteration of the statues creates feedback loops of the Serresean tales on various statues of our cultural making. They act as iterated input for the Serresean system we call "discourse on our epistēmē" in view of a new knowledge. But in the "seed story," the space of the "constant and close reach" between the canine corpse-object and the astronaut-subject allegorically restricts the range of motion of this system. Within these limits, knowledge binds object to subject in the company of death, but stretches out over an "astronomically infinite distance." For Serres, the allegory becomes a visual when he peers through the literal picture window in the middle of the writing and sees a certain region where death, object, and subject are linked. We say that the trajectory of Serres's understanding of the scene remains within this bounded region of fuzzy borderlines that blur the distinction between subject, object, and death: we must not forget that Serres cannot say with certainty whether the projectile flying alongside the rocket is defined as a subject (pet, man's best friend), object (frozen rock), or death (signified by a cadaver). In the same way, we must not forget that he describes the astronaut by what may be called a "patterned definition" with fuzzy borders: as a subject forming the core substance of an object (the rocket), the two becoming one system, bound together by a high-risk undertaking (the risk being death). But greater precision cannot be had in either case. The movement of the epistemological system from Verne's story to Serres's version of the same story is unpredictable *locally*; only *as a pattern* do we see knowledge emerge. Serres models the region of this knowledge as a statue because, anthropologically speaking, statues are a cultural invention that originally betrayed most overtly the signification system we are concerned with: an object of stone was shaped into a statuary placed on a burial site for the purpose of recalling the memory of a human subject taken away by death. Serres extends this anthropological origin of the statuary to the statue as such and models it as a dynamical epistemic figure whose significance can be fully understood only today, because it coincides with the concept of strange attractors of complex system domains. He recognizes that the original function of the statuary is uncovered by Jules Verne's narrative. The configuration of astronaut, object (first the stars, then the liveless dog), and a frozen projectile that recalls the event of death is a narrative statue. Helped by an insightful illustration, Serres's reading lays bare the dynamics of this statue. What is of greater importance, however, is the awakening understanding that within this configuration, death is an element which introduces undecidability into the system. Serres sees in Verne's nineteenth-century storied statue a postmodern simulation of a chaotic system which provides a pattern of knowledge with undecidable or fuzzy edges. This he then models in the remainder of his text as epistemological statues harboring nonlinear properties.

Serres's method is iterative. The Vernean canine corpse is a "statue" linking object (frozen rock), subject (Satellite, the astronaut's cherished pet), and death. Definable, separate terms in and by themselves, together they evolve into a projectile-like statue where their individual characters are no longer clearly discerned. Verne's reader observes the scene from the next vantage point: there is the object (the "thing" the astronaut observes outside his porthole), the subject (the astronaut), and death (indicated metonymically both in the canine corpse and the astronaut's mesmerized attention). Starting with this fictional setting, Serres's reading traces the statue as a strange and fuzzy pattern that now allows him to see the relationship between subject and object as he had never been able to see it before in his constant search for some form of union between the human being and the world of objects: "Death stabilizes the relationship between subjects and objects" (45).[20] Yet, death is an event that we normally consider the most destabilizing force of all human experiences. With the "new knowledge," our epistēmē should be redefined, no longer starting from a teleological perspective which gives rise to the construction of the idol Baal and the space shuttle *Challenger* for the express purpose of gaining sacred or scientific knowledge, but arising from an archeological effort to uncover the origins where the epistemologist hopes to meet death as a global generatrix. This is a process that asks for a descent into the statues littering the world, a descent that goes to their very founding principle.

The frozen cadaver reveals that a statue is not a simple form expressing stasis and fixity. As Serres looks closer at this simulation of a cultural and epistemological construct, he recognizes that the three "meaning-dimensions" formed by the dynamics of the subject, the object, and death, respectively, are really layers of meanings, and that, when he "zooms" in closer, each layer reveals its own layered composition, and so on. In his narrative account of seeing Verne's story, Serres expresses this by his futile efforts to state with certainty whether the "something" flying in formation with the rocket is subject, object, or a sign of death. He finds it as impossible to do when he attempts to gather the precise meaning attached to the image of the astronaut sitting in his object-shell. Meanings have become fuzzy or fractured. The storied statue is a space whose three-dimensional construct hides a system of fractal dimensions where the signification of subject, object, and death can never be clearly delineated. At each signification border a new space opens up that reveals the same form in scale-independent symmetry. In Serresean language, this is called a "descent" into unknown regions of the statue's foundation. I pointed out one such borderline pattern, where Serres as reader looks through the walls of the book and observes a dynamical statue, a statue that itself consists of an observer-object gazing at a similar statue which, in turn, cannot be dissected into locally precise elements but can only be implied as a *dynamical pattern*. The system is self-similar. The discursive statues that are presented in the subsequent chapters of *Statues* reveal these fractal dimensions of the system statue.

Once the nonlinear character of the initial scene has been described, Serres establishes a space-time context for a recursive series of statues. This forms the content of *Statues*. We may understand each discursive statue within the text as if it had functioned, for a given cultural construct before the age of computers, in the same way that today a phase-space portrait on a graphics screen models the mathematical formalism of a given complex system. Both simulate strange attractors that model global signification at a cost of local predictability. Harboring generative death at its innermost being, that is, as the reason why human beings fashioned statue-like forms in the first place, Serres's discursive statue models a strange attractor fractally traced out over time by the trajectories that describe the evolution of knowledge. No mere assemblage of statues or statue-like constructs, *Statues* iterates their evolution over time (their functions within the cultural context), using the complex system "subject-object-death" as the dynamic. This continuous iterative process, where output of one "statue-function" becomes input for the next, produces the discursive system's evolution toward a pattern of knowledge with fuzzy borderlines: Instead of predictable states of knowledge, the text dissolves into the fuzzy boundaries of a date-less epistemological soup where Serres finds the genealogical roots—not the tree—of knowledge.

The two levels of *Statues*, narrative and discursive, should not be confused: The chronological distance between the dated tales spun around each of the statues increases rapidly, until the regressing historical time scale becomes astronomically large. We know now that this knowledge based on historical time can never be precisely defined, but only globally outlined. It is expressed by Serres in the indeterminate "Without Date" (Sans date), title of the text's last chapter. Discursively, however, Serres's text arrives at a manifestation of time as the frame for a complex system in evolution. In fact, this chapter presents three statues in the form of mythological stories whose functions combine to mark a region of unpredictable states, time being the vehicle carrying knowledge along this region: global knowledge without local predictable data is an expression of a system determined by its own internal time, ahistorical time. Epistemological awareness of this time as a present-absent state unto which all our knowledge falls, is expressed by Serres as "new knowledge." Just as the strange attractor is never predictably present in a given complex system's evolution, yet determines its future states, the "new" time is not a locally calculable part of our epistēmē, as is so erroneously assumed for time understood in historical context. But it does determine all future epistemological states.

Serres's discursive statues must be read with great care so that they will yield the new time, for it is the silent companion of historical time frequently evoked in *Statues* and at times in beguilingly simple form. Two examples suffice. For one, the triadic "Table of Contents" is a historical construct; but its chronological inversion, recalling a similar inversion of Babel in *Genesis*, leads to a nonhistorical, mythological time and alludes to an epistemological upheaval in temporal perception. In a very different mode, Serres describes one of his childhood memories, a statue composed of bull and bullfighter. A figure of pure stability, it models the passage of Girardian substitution from human to animal sacrifices (81–85). It partakes, therefore, of the rock on which the resolutions of sacrificial crises are inscribed and history is founded. It documents the "statutory" time of the hominization process. Serres uses the linguistic link of the "statutory" to the statue, to simulate once again the fetish adoration in the admired figure in the bullring. It models one of three curses that regenerate history through multiple murderous foundations and produce the forgetfulness that looms so large in the early pages of *Statues*. Yet, the reader is left with the same allusive indecisiveness that perplexes Serres: how to tell the full meaning of a statue that is chaotic (85). Left unanswered, and therefore undecided, is the definition of historical time as knowledge through language. The statue as visible form of the statutory, of the law, is also the visible form of language, the expressive form that holds Serres's attention throughout his discourse in *Les cinq sens*. His question states the impossibility of the statue, as a paradigm of pure stability and language, *to go beyond*; the question betrays a fascination with the forgotten generatrix, what Verne's story expresses through death as a stabilizer.

The step beyond, Serres calls it a "descent through the statue," through its socle, past its rocky foundation, into the netherworld and a face-to-face encounter with death, told on the back-cover of *Statues*. This "spiral" movement is the discursive statue of *Statues*; conceptually, it goes far beyond the statue that we see only as a place (*lieu*) in our forgetful and delusionary thinking; it goes beyond any linguistic representation. Instead of a place of deceptive stability, *Statues* is the story of a dynamical statue, a "structure of change and movement," a discursive expression of the "new time." The story of Sisyphus serves as an example of the silent time. With one of the rare footnotes in this text, Serres recalls his own reading of the myths of Sisyphus and the Danaides, in *Hermès IV: La Distribution*: It demonstrated that stability, even expressed as an eternal return, is a special case, a singularity. Entropy and the irreversibility of time, on the other hand, are the general condition and the most common deviations (*inclinaisons*) from stability, without which production and renewal of life and nature would be impossible (*Hermès IV* 219–225). But this is a reading that concerned science, and natural or physical time, not historical time. In its epistemological perspective, *Statues* presents the same myth but reads it very differently. First of all, writes Serres,

> the various interpretations of the myth, including my own [in *Hermès IV*] with its scientific calculations, speak of nothing but the scene and its hero who has become the guilty, wretched slave-laborer. We habitually see only ourselves. . . . And what if, for once, we were to look at the rock, that unchanging presence before our eyes, that stubborn object resting in front of us?[21]

Serres extracts the object from the obscurity epistemology relegates it to while analyzing Sisyphus' fate. He tells us about Sisyphus' forgotten ancestors, survivors of the original flood who understood, as yet another myth relates, that the stones they threw out of the ark and over their shoulders were the bones of their parent; or, which is the same, that the bones of their parent they had gathered into the ark were the stones of the earth. These survivors of the flood understood that the object was at the same time the subject, that the "at-the-same-time" was made possible through death, and, what is even more important, that from this understanding and their corresponding action were born men and women, and post-diluvian life began (304).

There is a striking similarity between the function of the stones in the myth and the function of the frozen dog in Verne's tale. Sisyphus is the descendant of his ancestors' life-giving action, he is not a mere slave-laborer:

> We had lost the memory of the stone being moved [by Sisyphus]; we had forgotten that it was at once our immemorial past and our total future, that is, the human race. Stones form the skeletons of the dead and the seeds for the future.[22]

This wonderful reading turns the myth of Sisyphus into the tale of a statue-like combination of subject-object-death, and links it to an ahistorical time that is neither physical nor historical. It is the model of a dynamical system that cannot forget its reference to time. Here, the "collective immortality" receives its simulation in the dynamical discourse that Serres spins around the Sisyphus myth: The king, condemned to retrieve a stone from the lower realms, could as well be retrieving and thereby rejuvenating the bones of his ancestors. The stone-bone rolling down the mountain has the same trajectory of meaning as the stone-bone tossed over the shoulder of his ancestors.

In *Detachment*, immortality is modeled through fables, first with the image of a Californian sequoia tree, then by the countless inventions that drive human culture. In *Statues*, Serres replaces the apologues and anecdotal evidence with a dynamical system where subject, object, and death are related to each other, by rules that mythology, including Verne's, spells out for those who know how to read without forgetfulness. This epistemological system evolves toward a global awareness of the triadic relationship sustaining a life-giving pattern. But locally chaotic uncertainties reign, expressed by the logically pointless and futile Sisyphean task. Although we commonly shudder at its endlessness, and remain in awed or horrified awareness of time's senseless and therefore chaotic nature, time is, in reality, the motivator or condition for this ambiguous but vital knowledge. This is the lesson of the Sisyphus myth. Serres calls the acquired knowledge "collective immortality;" in nonlinear terms, we call it "ahistorical" or "dynamical time." No matter by what expression one refers to it, its epistemo-critical mode is "implication" as opposed to explication (the Ulyssean exode instead of the Cartesian method, as presented in *Les cinq sens*). Implication is the epistemologist's reference to a nonlinear approach that dynamicists call "global signification" with local unpredictability. This odd knowledge is simulated by Serres's epistemological version of a mathematical strange attractor.

Before we put *Statues* aside, we have to look at one more statue, since it justifies our initial proposition that Serres's discourse itself is nonlinearly constructed and harbors a strange attractor. He compares his own work to that of Sisyphus and gives us a cursive summary, of sorts, of his writings—modeled on the statue (309–311). Since *Hermès I*, he writes, the statue has always been in his discourse, a heavy mass of granite, a massive block of stability, accompanied by the threat of stagnancy and death (the figure of the Molièresque Commander being one of the earliest examples). Like Sisyphus, he thought he was laboring valiantly but was engaged in repetitive gestures. No doubt referring to himself, Serres can now state: "Toiling without reprieve, under the constant presence of the invariant [the statue], this work finally has arrived at a stage, here in this text [*Statues*], which allows for the tracing of the trajectory of the Sisyphean rock no one ever talked about" (310).[23] The discovery of the true meaning of the "Sisyphus-Serres" myth is his admission that the static statue, so haunting in his earlier writings, has become a dynamical statue, with all the liberating inventiveness the nonlinearity of this "tracing" or dynamical construct implies.

Statues presents the function of Sisyphus' rock, through a dynamical reading of statues that model how all facets of our knowledge tend toward bounded or orderly chaos, a truly strange attractor expressed through implications of the "collective immortality." Serres insists further that this convergence has always been visible for those willing to look. Though consistently excluded by and from scientific thought and cultural consciousness, it was expressed through other phenomena whose locally diverse characters the Serresean text outlines in order to arrive at their common founding in death. This permits him to unite them under the singular, aesthetic sign of the statue. If one reads the Serresean statue like a discursive strange attractor and follows the philosophical implications of the latter for modern science, one understands why, in the very last pages of *Statues*, the dynamical statue leads to the epistemo-critical question of space (*le lieu*). It evokes a quest, not for the statue, but for the "fuzzy space" of the evolving dynamical process that the statue inhabits. We may call it "time," a time derived from the dynamical process. Serres's answer is immediate and perhaps the most visible proof that our reading based on complex dynamics is eminently appropriate:

> Chaos can be defined as variability where disorder prevents the appearance of a place (*lieu*). Chaos does not offer any reference points, or, perhaps, we cannot discern any. . . . Chaos does not respond to the question "where?"[24]

Instead of demarcating a predetermined space of knowledge, it implies only—but this is everything— a region or basin of unpredictability. Whereas stasis occupied the entire epistemological space (allegorized in Alexander of *Detachment*), post-modern epistemo-critical approaches concern the unanswerable question about the place the statue has marked for so long as the space of stasis. But simulated through what we have read as a dynamical statue, or a discursive attractor, the unanswerable (or "unknowable" of Weissert's terminology) is endowed with global signification constructed along ahistorical time.

For Serres, stasis has changed into an interrogative expression of place marked by a dynamical complexity we can understand only with reference to time. It constitutes a paradigmatic shift that parallels the epistemic shift to strange attractors in mathematical-scientific thinking. The impact of this shift and the fundamental switch it forces on contemporary science is very succinctly described by Mitchell Feigenbaum, mathematical physicist known for his formulation of the "universality theory" which offers a quantitative understanding of the onset of turbulence in nonlinear dynamical systems:

> The whole tradition of physics is that you isolate the mechanisms [of a system] and then all the rest flows. . . . That's completely falling apart [today]. Here you know the right equations but they're just not helpful. You add up all the microscopic pieces and you find that you cannot extend them to the long term. They're not what's important in the problem. It completely changes what it means to *know* something. [emphasis is Gleick's] (quoted in Gleick, 174–75)

By investing the statue with non-periodic dynamical traits, Serres points to a qualitative change in the postmodern epistemological perspective. *Statues* demonstrates that the history of human epistemic construction has always had in view a *different* "knowing-something," and that it left clues to that effect in the form of statues. Our task is to acquire this "other" knowledge, in a process similar to the way in which it is becoming acceptable to scientific thinking today.

5

Time Inventive: Reading *Le Tiers-Instruit*[1]

Time is invention, or
it is nothing at all.
—Henri Bergson

In *Statues*, the object of Serres's investigation is the development of each of the statues discussed as a dynamical system. The text's logical task is therefore to pry each statue loose from its stationary mooring which is Serres's figurative formula for deadly stagnancy. In the previous chapter, I endeavored to show that the difficult process of detaching true knowledge from its static socle of deadly abstractions (analogous to Orpheus' task of recomposing Eurydice's concrete and "true" body) implies a new mode of thinking or knowing death. No longer a destructive force, death functions in *Statues* as Serres's primary trope for a creative force that is part of the dynamical system the "new" statue simulates and that contributes to its nonlinear complexity. In its essence the statue is presented as a "timing of space." This portrait may be read as one that finally comes to terms with Serres's often critical commentary on Western thought to which *Rome: The Book of Foundations* lends a sometimes bitter note. The Roman statue is the overriding structural trope in this text, but in a unitary configuration where death is posited in a teleological sense as a corollary for all power structures developing into monolithic phenomena of human historical experience. With the exception of its last chapter (see introd.), *Rome* remains Serres's discursive plotting of a *conservative* dynamical system that "describes the world in terms of deterministic and reversible trajectories" (Prigogine and Stengers, 39).

Statues's discourse allows for a different understanding of death. This is obliquely preformulated in the text's subtitle, "Le second livre des fondations." Marking its retrospective role in regards to *Rome*, *Statues* inscribes a "timing" process that functions also in a projective mode. By arguing for what is in essence a *generative* role assigned to death, it offers an initiation into the dynamical process that effectively serves as an underpinning for *Le Tiers-Instruit*, which appeared in print four years later. This text develops a story of the "tiers-instruit" on a level that resembles what Alexander J. Argyros calls "epigenetic rules." These express "a combination of genetic and cultural transmission" which "can be thought of as interstices between biology and culture" (Argyros, 219, 221). They do not in themselves constitute the character traits of the human being but "must be thought of as guides or head starts"; they are "a kind of systemic memory that serves as background for productive change" (Argyros, 302, 303). In this sense, *Le Tiers-Instruit* is the blue print of a dissipative dynamical system.

Le Tiers-Instruit is a book on education in the largest possible sense of this term: the all-encompassing formation of human thought as *invention*; and vice versa, inventive creativity as the only activity worthy of cognitive thought. Indeed, Serres's text can be read as an elaboration of Argyros's dynamical "formula" for beauty which combines evolution, chaotic self-organization, art, and invention as "a beautiful disruption of ecological equilibrium" (Argyros 287). "It's a narrative" (*Conversations on Science, Culture, and Time*, 161) of one particular statue, namely, the human as such. In *Les cinq sens*, Serres connects his fables directly with nonlinear dynamical systems theory, when he compares Ulyssean meandering and the philosophy of circumstances with fractal basins which hold islands of inventive knowledge emerging like strange attractors out of chaotic fluctuations (*Les cinq sens*, 287). *Statues* implies this connection in its attempt to redefine historical foundations in more dynamical terms or "second foundations." *Le Tiers-Instruit* discourses in the same language, but its concepts are couched in a theatrical setting: the Harlequin-prologue. As in the case of *Genesis*, this prologue, too, represents all that follows in *Le Tiers-Instruit* and serves to give the "instructed-middle" or the "middle-instructed" (*le tiers-instruit*) a name and a face before we begin to follow the author on the difficult and chaotic journey into the topology of human upbringing.[2]

In *Statues*, the statue as the primary trope for a discursive strange attractor is given a final outline, when Serres brings his discussion to a close with an "androgynous Hermes." He is the mythological equivalent of structured chaos or "chaos bound" (Hayles). In *Le Tiers-Instruit* the mythological reference is replaced by a cultural one personified in Harlequin and his twin Pierrot, both traditional figures of the French theater. The choice of this French practical joker is highly significant. But Serres gives no hint of this, since his primary reader is presumably French or francophone familiar with French cultural heritage. The English-speaking reader should see in the prologue a deeply imbedded cultural allegory, in order to fully appreciate the global implications of the "instructed-middle" presented on the narrative stage of *Le Tiers-Instruit*. We must make a qualitative leap to and into a new way of thinking that is implied in this choice.

In the same way that the circus clown has become an American institution with his familiar appearance (the red nose, the big smile outlined by grotesquely overblown lips, the oversized shoes) and a certain standard repertoire of performing action, Harlequin is a French figure with a long tradition going back to the Italian commedia dell'arte of the sixteenth century. Popularized in France in the seventeenth century by the actors Biancolelli and Gherardi, perhaps the most famous of all the improvisations required by this theatrical art form was Harlequin's role in *Arlequin Empereur dans la lune*.[3] Only after particular improvisations became fixed in the imagination of the public, were they much later committed to writing, which explains the great variety of versions of these comic skits.

Under Serres's pen, Harlequin presents himself once more in a novel improvisation and becomes a stage model for a complex dynamical system whose behavior over time displays chaotic and ordering tendencies. This strategy is two-fold and almost contradictory, yet typical for much of Serres's discursive-logical style: He honors an old theatrical tradition of improvising on the behavior of a stock character who must always appear in the same costume of a patchy multicolored garment, with a black half-mask and a wooden sword; what he implies, however, is postmodern science at the cutting edge, explicitly outlining concepts of chaos theory.[4] Staging the traditional Harlequin is Serres's excuse for a stage production starring fractal dimensions, strange attractors, and dissipative, self-organizing systems. Two seemingly incongruous modeling systems are juxtaposed, although they are far apart both in historical time and fields of human experience. A late-twentieth-century dissipative model that will give us a better understanding of the chaotic systems of our world is introduced by the traditional French version of a harlequin.[5]

It was customary in the sixteenth- and seventeenth-century theater to affix brief outlines of a given scenario in the wings of the stage as a sort of prompt or cue card for the actors to peruse during their improvisational acting out (Storey, 6). Serres "tacks up" his harlequin story on the prologue wings of his textual stage before presenting the education of the "tiers-instruit." In the course of a discursive improvisation that fits into postmodern thinking, Harlequin stages a scenario of the relationship between the local and the global. He acts out what is expressed on the scientific level between classicists and pioneers of dynamical systems theory.

In his role as emperor of the moon, Harlequin is riled by his audience for insisting that his lunar voyage did not reveal anything new and that "everywhere everything is the same as on earth" (11).[6] Serres is deliberately vague when he refers alternately to a press conference and a lecture hall full of eager students who are disappointed by Harlequin's explanations.[7] The hapless clown decides to remove his coat the very appearance of which belies his words. It is a vestment composed of tatters so wildly varying in colors and so chaotically arranged that the resulting quilt resembles a crazy map puzzle where nothing "here" is identical to "something" there. For us spectators, the tattered mantle is the model or discursive visual of a complex system. Harlequin, however, insists he can explain his appearance and bring it under control or to unitary, orderly sameness as soon as he takes off the offending vestment and exhibits his royal self underneath, like a paradigm of truth. That is to say, his explanation will be of a perfectly linear nature, resembling a linear equation, with an explicit, fixed-point solution.[8] But removing the cape only reveals another equally tattered and color-strewn garment below. This means that the transformation operated on the initial state of the system in order to evaluate it (that is, removing the chaotic coat and finding a clear answer underneath) generates a new but equally random state which, when evaluated in turn (the second coat removed), allows for the production (or appearance) of yet another indeterminable state (or ragged garment), and so on to his tattooed skin.

Harlequin's striptease is Serres's portrait-model of a series of functional iterations operated on the initial phase of the system (the first cape enveloping Harlequin). For his part, the insolent pupil who was the first of the spectators to wonder aloud about the inconsistencies between Harlequin's words and his appearance, is the student-scientist who recognizes a highly complex nonlinear system. He knows that no fixed-point attractor or limit-cycle attractor would ever offer a solution. A model of a fixed-point attractor or satisfactory answer would appear if, in fact, Harlequin could undress to his bare royal skin, exposing an unmarked expanse of sameness. Such a result would give credence to the disappointing results of his exploratory findings: The same laws apply always everywhere. They would compose an equilibrium state guaranteed by the royal "equilibrium" as keeper and guardian of the truth. A limit or periodic attractor would be the appropriate model, if Harlequin's performance were to become periodic: in this case, every unveiling would present something identical to the initial state of a tattered, color-strewn surface (garment or skin), forever—or as long as the audience finds it amusing, before tiring of the repetitions and simply halting the performance.

But neither scenario happens in Serres's improvised staging: "Even Harlequin's [tattooed] skin belies his assurances that unity is a global law" (14).[9] The horrified audience finally realizes, a little slower than the rebellious pupil, that skinning poor Harlequin would not do, because he has no fixed point and cannot be defined in binary terms or periodically recurrent settings. They discover that he is androgynous, man and woman, human and beast, emperor and slave, neither young nor old, or both simultaneously; he is monstrous in his endless, unpredictable multiplicities. His very being as a *corps mêlé* will *forever* iterate the initial number, that is, the question of sameness and otherness, without ever arriving at a fixed or periodically recurring solution. We are again reading the figure that complex dynamics calls a "strange attractor." The question who Harlequin is, what he is, where he is, cannot be resolved with precision. Under Serres's improvisation he has become a model for fractal dimensions. He is a fractal: Harlequin is there and is not there, he answers a lot of questions, but every answer is another question in a self-similar movement. He is Serres's epistemological model for a strange attractor.

On a computer screen the strange attractor can be traced out as a basin toward which the trajectory of a dynamical system's orbit converges while looping ergodically through its phase space. In the text, this looping is discursively portrayed by the harlequinesque *métis*, a chaotic body or half-breed who goes on halving himself like a Cantor set or a Koch curve.[10] The entire scene is fractal. Suddenly the chaotic confusion of shapes, colors and other attributes which compose the hapless Harlequin on stage, is transformed—or so the audience perceives it: In the place of Harlequin stands something luminous, which Serres describes in terms identical to those he used to describe Gillette (17). It is Pierrot, another famous clown figure of traditional French pantomime, who is always dressed in white with a white-powdered face, void of any distinguishing features, chaotic or orderly. Having turned away from the stage in disgust and ennui, no one in the audience knows how many iterations (or peelings) have brought Harlequin to this phase. All one can say is that the nth iterate of the transformation on the system "Harlequin" is no longer identifiable; it is a white statue that seems to blur or melt into blinding luminosity. Locally recognizable differences defined by color, contour, form, and/or texture, and randomly agglomerated into the chaotic layers that compose Harlequin, are replaced in Pierrot by global recognition, but at the price of precision and exactness. The blurring of forms and the blinding luminosity that together are "perceived" by the audience as the white Pierrot can no longer be described in more precise terms. Pierrot is an image of a strange attractor. Old cultural forms and cutting-edge science are once again wedded in Serres's text.

Readers who know Serres's earlier works are familiar with his frequent use of whiteness as a trope. One need only think of *Rome* where Rome's rival city Alba and its history become the allegorical background for his exposition on whiteness or blankness (both expressed in French by one word, *blanc*) as a model for the paradoxical concept of "all *and* nothing." In *Genesis* we read the same in a more abstract formula: "Blank, undetermined—POSITIVE" (48). In *Le Tiers-Instruit*, the traditional white costume and powdered face of Pierrot serve to guide the reader to an understanding of a strange and unorthodox concept: the all-and-nothing, the excluded middle un-excluded (not merely included), renamed the instructed-middle. Serres's technique is used by all good teachers who use a common, well known object (in this case Pierrot) to explain a difficult, philosophically new concept. He will later call it the "middle philosophy," (*la tierce philosophie*).

The choice of Pierrot and Serres's almost exclusive emphasis on his blank, transparent, and fuzzy appearance hint at a fine understanding of the strange attractor and the role this concept plays in nonlinear dynamics. This buzzword of chaos theory is often transformed into something it is definitely not, as Thomas Weissert emphatically explains:

> In classical mechanics, the point attractor is present at one point in phase space [= signification space]. Not only does it determine the future of all trajectories, but it is also "always already"; it is identical to that future. But [in a complex dynamical system] the strange attractor breaks away from this superposition; it is no longer simply identifiable. Recently, we have seen

> many pictures of so-called strange attractors, and it is true that one thinks of these patterns when strange attractors are mentioned; but these pictures do not capture the strange attractor. A trajectory on a strange attractor is merely a *trace* of the attractor. [emphasis is Weissert's] The attractor is the object that the trajectory would trace out if we could allow it to evolve for an infinite amount of time. The sites that together comprise the attractor may be bounded or enclosed within a finite region of phase space, but they are strung together in their peculiar manner only in infinite syneresis. The attractor is not present in phase space. . . . Still, the attractor must be present in some sense because it determines each event—each subsequent state of the system presence. (Weissert 1995, 127–128)

Serres's improvisation of Pierrot is a lesson by theatrical means on the way Weissert urges us *not* to conceptualize a strange attractor. Harlequin is the embodiment of a complex dynamical system whose behavior over time we can see as a graph if we choose to plot it in its phase space, and as a cultural representation if we choose to plot a stage production. As different as these two processes seem to us, issuing from the opposing domains of science and the arts, from the standpoint of a dynamical reading they are "isotropic"; they display features specific to similarities in observed behavior-over-time of both scientific and cultural systems (Weissert 1995, 113–14). This allows us to say that Pierrot is the disembodied portrait of a strange attractor which cannot be captured in a picture or graph. In the text, the two figures find their nexus: Harlequin is the discursive trajectory that "traces" the attractor as it tends toward the attractor basin "Pierrot." Together, Harlequin and Pierrot represent the traceable *and* unidentifiable of a truly mixed body, soon thereafter called "l'île nulle de mémoire" (49).[11] One of the first simulated strange attractors in *Le Tiers-Instruit*, this zero island of memory models the instructed-middle.

The Harlequin prelude bears the title "Laïcité" and ends with a mise-en-scène of the "secular miracle of tolerance" allegorized in Pierrot. It is a dramatization of the epistemological laicism Serres proposes in *Les cinq sens* as a necessary conditon for "new" inventive knowledge. The laic miracle erases the boundaries between the sacred and the profane, between the initiated and the excluded. Leading to genuine freedom of invention and thought, it is presented in the text proper of *Le Tiers-Instruit* as the objective for an apprenticeship in the ways of the instructed-middle. Following the textual staging of the prologue, the first part of *Le Tiers-Instruit* called "Elever" describes improvisations of Harlequin as a human being on the contemporary stage of our world. Here, the principal paradigm remains faithful to what the joker stands for: a third element that is neither this nor that, neither here nor there; it is both and neither. Whether Serres talks about the left-handed child (himself) taught to write like a right-handed child, the swimmer who arrives at the middle of a stream where he enters a space that is neutral in distance to both shorelines, or the goalkeeper who must move in all directions and none to guard his goal, all are harlequinesque in their dilemma: they arrive at a point where commonly accepted definitions fail, where "all directions and meanings are equally valid" (27).[12] To learn to live or pass in this "middle-ground" (*la tierce place*) is what Serres calls an "apprenticeship for the making of the third." It changes the excluded middle into an instructed-middle.

The truly instructed-middle leaves all differences behind by combining them all (24): The hyphen of the "tiers-instruit" stands for the sliding or passing middle fold or ground, neither stationary nor moving, where the apprentice learns to live in the middle of differences without differentiation. This is more explicit in French where "hyphen" is rendered by the expression *trait d'union* (a line bar of union). Serres stresses its importance when he characterizes *Le Tiers-Instruit* as "a narative [which] has as its theme a meditation on the bond [*le lien*]" (*Conversations*, 161). It is of interest to note in this context that in the first Latin texts to be printed in the fifteenth century, an inverted capitalized omega was used to connect two words, later changed into the now familiar hyphen (Grevisse, *Le Bon Usage*, 61). If we read the hyphen, therefore, as a distorted or worn version of an *upside-down* omega or end, used to finalize a union, it can be understood as operating in *Le Tiers-Instruit* in a fashion that resembles the "inverted" Babel of *Genesis*: it finalizes a union, yet its inversion prevents a static rest or end in itself. Seen in this light, the hyphen imprints diacritically the meaning of the "middle-ground" (*tierce place* or *tiers-lieu*) (34). Neither posing (thesis) nor opposing (antithesis), the exposing middle (or third) is the same as the "non-present present" by which Weissert characterized the never identifiable status of the strange attractor.

The portrait of the instructed-third is informed by a learning process resembling a voyage, motion, bifurcations, becoming, and phase changes. Being a learner means becoming manifold and bifurcating:

> No, the pedagogical game is not played in two's, traveler and destination, but in three's. Somewhere, the middle-place intervenes as a kind of threshold for the passage. This door, most of the time neither apprentice nor teacher know its place or its use. . . . Then one day, there is the middle of the river, this strange state of phase transition that one could call sensitivity, a word which means possibility or capacity for all directions (all meanings).[13]

The "tiers-instruit" is the human child Serres describes on its way to an adulthood of invention and thought, in terms that function parallel to dynamical systems concepts. The analogy drawn between the scientific concept and a new pedagogical approach to the education of a middle-instructed human being serves to unite the text's sometimes bewildering array of seemingly disjointed examples, stories, and anecdotal evidence into a commonly shared model. This strategy facilitates its reading without destroying its stylistic dynamics. In an oddly reiterative way that confirms Latour's observation that Serres talks about the "known" rather than the "knowing," the reading of *Le Tiers-Instruit* requires, in fact, that one has already gone through the apprenticeship of the instructed-third. Reading dynamically with nonlinear concepts and with a careful reflection on the Harlequin-prologue, this is not too difficult a venture.

The argument of the text follows a logic that itself resembles a dynamical system: Serres presents three major tropes at the onset of the passage to the middle-place; they represent the rules, or the dynamics, by which the system called the "learning child" evolves. In mathematical language, we speak of equations; here they are tropes: "becoming manifold" (*devenir plusieurs*) for which the left-handed child learning to become ambidextrous is the concrete example; "facing the outside world" for which the swimmer is the illustrative example; and "going in all directions without excluding any" (*bifurquer ailleurs*) of the goalkeeper. In Serresean logic, the three tropes determine how the system "human child" moves invariably to the "third" or middle-place, the region where one arrives at knowledge without ever fully grasping it (31). At various moments of his argument, Serres summarizes these three rules or tropes (29, 31, 32, 33, 39), at each repetition their functional values slightly shift or progress, and our understanding progresses at the same pace. We may term this as symptomatic of iterations on the behavior of the system defined by the three rules.

Recognizing the middle-place in these arguments implies a fuzzy region of knowledge that the system "human being" tends towards on its voyage of apprenticeship. The definitions, so precise at the outset by virtue of concrete examples of everyday life, become lost; that is, they no longer produce *expected* results at the threshold of the middle ground or place, analogous to the behavior of a nonlinear dynamical system that evolves over time into unpredictable states. However, what Serres discovers there is a globally recognizable "strange" region of inventive possibilities. Here, analogous to the basin of a strange attractor with fractal dimensions, the instructed-middle lives in a time that is a "pure passing," precisely in a "timing of space":

> All evolution and all apprenticeship require the passage through the third place. In this way knowledge, as thought or invention, never ceases to pass as a third through the third place, [it is] always exposing itself; and he who knows, thinks, or invents becomes very quickly a passing third (un tiers passant). Neither posed nor opposed, but forever exposed. . . . Here then is the almost complete description of the instructed-middle [le tiers-instruit] whose instruction never stops: by nature and through his experiences, he has entered into time.[14]

But Serres adds that he *does not dwell* in time. This echoes the description of a strange attractor traced out in phase-space by the trajectory of a complex system as it evolves over time yet never arrives or comes to rest in the attracting region. In other words, the strange attractor is traced out by the *continuously* varying evolution of a system, for which the instructed-middle stands as a Serresean simulation.

The section "Elever" is thus a description of the Harlequin-philosophy (*la tierce philosophie*) useful for the upbringing of the human being (56). In fractal style, other tropes continue the demonstration without ever quite repeating the exact argument of previous ones. One could briefly summarize this section by recognizing a progression in three stages: the tropes of the body (the ambidextrous child, the swimmer, the goalkeeper); the analogy with the soul as an intimate and personalized image of the middle-place (65), which plays a significant iterative role in the last section of *Le Tiers-Instruit*; and finally the author who offers himself as a trope for the "instructed-middle" in the passing of the "middle place" (61–65). Perhaps the most enigmatic of the three parts that compose *Le Tiers-Instruit*, "Elever" is the easiest to enter into, provided one has understood the underlying structure of a complex dynamical system both on the content level and in the discursive style. Merging style and content, the complexity of which is inherent by virtue of the dynamical approach, Serres now says: "Harlequin becomes Pierrot".[15] This powerful yet elegantly simple formula is Serres's model for a pedagogical blue print: A kind of "human" strange attractor evolves like a dissipative system into a region of unpredictable luminosity. Its ultimate goal is to achieve tolerance which does not exclude (Serres's *laïcité*), and inventive thought which acts as the passage to a new knowledge. Already in *Rome* Serres had given a simple but compelling reason for such an upbringing: "I want to remain free, an instructed third" (173). The question of intellectual freedom as the basis for true learning becomes here ontological reality.

From the viewpoint of content, the second part titled "Instruire" is perhaps the more practical of the three parts that make up *Le Tiers-Instruit*. The human being raised into a middle-place, object of discussion of the first section, now enters the actual schooling in the ways of a knowledge that sheds all vestiges of epistemo-centrism or totalization (70). In order to tell this story effectively, Serres introduces Johannes Kepler's elliptical orbits of the planets, and the problem of the diagonal of a square with side of unit 1, a problem that ancient Greek mathematicians struggled with. Both examples, astronomical and mathematical, are superb metaphors, if not definitions, of the middle-ground or the instructed-middle. The first speaks of Earth as the object moving on an ellipse of which the sun is one of two foci, while the other focus and the actual center of the ellipse are given over to oblivion. By reason of their obscurity, they constitute the "third place" and a lost middle zone (which here carries the same meaning). The second example demonstrates that, although the diagonal exists and is a useful geometric notion, it cannot be rationally measured. The square root of 2, representing the length of the diagonal, is an irrational number, that is, it cannot be expressed as a ratio of two integers. As an irrational it is "excluded," but within the new knowledge becomes a fertile middle-ground for the instruction of the instructed-middle (79).

Having entered this zone of excluded entities, the reader leaves behind the clear, sunlit, brilliant focus of reason and measurement, and gains a forbidding region of darkness and forgetfulness, the twilight zone between brilliant focus and obscure focus, the transparent fuzziness of Pierrot. The third or midpoint of the ellipse, on the axis between its two foci, is a center, says the mathematician. But for the astronomer that same center is engulfed by darkness when the sun is displaced to an elliptical focal point and blinds the "reasoned" center. We may perceive Serres's Harlequin as an allegorical dramatization of Kepler's discovery and as a staging of the dilemma of contradiction encountered by the early Greek mathematicians. At first king of the stage, center of all reasonable knowledge and truth, assured in his royal uniqueness as the One, Harlequin removes himself from center stage, step-by-step with every peeling, until nothing recognizable is left of him. He becomes the dark middle point between the reasonable stage scene of the emperor of the moon (Kepler's sun) and the obscure, incomprehensible exhibitionist of his own chaotic being (the second, dark, unspoken focus of Kepler's ellipse). As Pierrot, he has become the middle-ground, both luminescent and obscure-transparent, with all the colors of his spectral garment resolved into a fuzzy zone: he is one and multiple.

Serres has thus set the stage for an elliptical reading of our epistēmē where the "black-hole" center makes for unpredictable states along the reader's future reading/learning orbit. A mere descriptive listing of these states, expressed in the text by the most varied examples of human experience, would not do justice to the metaphorical demonstration how the middle-instruction works to guide a "tiers-instruit." They are best read in their original text, for the power of their poetic beauty and inventive imagination. They are the content of the second part "Instruire." It is, however, useful to emphasize two aspects of Serres's style which is, as always, intermingled with the content to form a middle-ground. First, Serres chooses from among the most common concepts and occurrences to the most obscure data of our epistēmē, quite often characteristically French examples, especially as regards literary references. These, he interprets, reads, and implies in a most uncommon, surprising and unexpected fashion. Occasionally, even after understanding has come, this makes for a reading that remains elliptical, both in the traditional and the Serresean sense. These hesitations are not failures, of course, but moments when understanding is passing in the middle-place, when the reader has become the swimmer at the instance (and instant) of that neutral zone in the middle of his river, when he is like the goalkeeper stretching out in all directions, on the verge of decision. This is part of the "middle instruction."

Second and directly related to the first aspect, Serres uses the result of each argument or exposition as the starting point or seed number for the next example, in truly iterative fashion. The resulting recursive form displays self-similarity and imparts a dynamical quality to the text as a whole which we may call "nonlinear," given the highly complex initial input. The reader is therefore well advised not to read "here" and "there," for, while enjoying the beauty of the tattered pieces, she will surely miss the inventive creativity of the multiple whole and thus fall short of an unpredictable but complete understanding. Only a dynamical reading that follows the topology of the text becomes a "middle-reading"—global aim of Serresean middle-instruction and "elliptical" understanding.

Statues, too, is written in this recursive style which resembles scaling properties that dynamicists speak of when they work with fractal dimensions. Understanding for a dynamical systems scientist is based on modeling reality from often quite mundane details, like cotton price variations, stock market swings, and water eddies. When James Gleick describes the history of this process, he relates the case of Benoît Mandelbrot who formed an unorthodox picture of reality by studying the unstable behavior of all types of natural and man-made phenomena. In this way

> "[t]he picture was beginning to come into focus. . . . His studies of irregular patterns in natural processes and his exploration of infinitely complex shapes had an intellectual intersection: a quality of *self-similarity*. [emphasis is Gleick's] Above all, fractal meant self-similar. Self-similarity is symmetry across scale. It implies recursion, pattern inside of pattern." (Gleick, 103)

In the context of *Le Tiers-Instruit*, it is equally important to emphasize Mandelbrot's *intuitive* refusal to reject these *irregular* patterns as garbled disorder: "Intuition is not something that is given. I've trained my intuition to accept as obvious shapes which were initially rejected as absurd, and I find everyone else can do the same" (Gleick, 102). Mandelbrot's intuition is Serres's middle-ground where, and from where, dynamical knowledge passes. The intellectual intersection Gleick speaks of is the region of Serres's "middle-instruction" where the unitary and the multiple become self-similar. In very much the same way, our reading of Serres's story and the concomitant exploration of complex dynamical systems models is the intellectual intersection of a dynamical reading; we find self-similarity or "symmetry across scale" in content and language (style, metaphoricity, lexicality, grammar).

What Serres presents in *Le Tiers-Instruit* is a further lesson in topology. *Detachment* and the section on seeing/visiting in *Les cinq sens* offer fables that metaphorically tell the story of topology, the study of surfaces, and how to transform them by manipulations called "mappings." In mathematics these surfaces are abstract planes; in the reading of Serres's dynamical discourse, they are "meaning spaces." The floating, dancing spaces of *Detachment* and the meandering/visiting of localities amid their circumstances in *Les cinq sens* (discussed in chapters 2 and 3 of the present study) are the spaces of epistemological meaning

Serres is critically confronting. The transformation of these fields of meaning, so that epistemo-critically new "properties" appear, is also told through stories of sailors, farmers, Ulysses, and others. With these allegorical figures he shows the mappings that allow for the desired changes of the meaning space.

Statues and *Le Tiers-Instruit* go a step further. The discursive figures now model not only the resulting transformation, but also the transforming process of nonlinear dynamics. The transformed dynamical meaning space becomes the space of a new, inventive knowledge. Serres invites his reader to follow him onto the intellectual intersection where a patient passing of experiences becomes an adventure in *understanding topology*. The conceptual notion of topological mapping or transformation of meaning space also supports the claim of the present study that concepts of nonlinear dynamics can be applied to Serres's epistemological studies on time. If such applications function as mappings of one meaning space onto another and produce new open-ended meaning spaces, then the question of metaphorical, analogous, or literal use of scientific concepts is of secondary importance. The essence of a dynamical reading is a "topological reading" where the reader should be prepared to follow every transformation of meaning or metaphor in a textual phase-space that models multidimensional or fractal knowledge.

A topological reading implies a topological transformation of the language itself, for which the analogy between "tiers exclu" and "tiers-instruit" is the primary example. The language of Serres's text itself acquires properties that resemble those of complex dynamical systems, the very ones this language tries to *model* in its discourse. This process is evident in the choice of key words, with preference given to those whose etymologies have various and often borderline connotations, semantic bifurcations deliberately left unresolved. The discussion of Kepler's ellipse with its dark and "missing" center, for example, leads Serres to the word *cento* (centon) which is a poem composed of fragments and verses borrowed from other poems, but can also mean a patchwork garment. The Greek root is *kentrôn* with its many bifurcating meanings including a prod, the victim of such prodding, or the place where the prodding occurs; but it also has the derived meaning of a fragmented poem (71–76). Of course Serres never equates the unitary "centrum" and the fragmentary patchwork that is "kentrôn," but makes an analogy that is modeled like a topological transformation. We begin to see in the "missing centers" a new region for the middle-ground, the obscure middle of an ellipse, precisely a *cento* in the figure of the chaotically colored Harlequin, victim of the humiliating prodding by the audience.[16] Another example is the whimsically suggested neologism "chaopedia" to replace encyclopedia so that "instruction by embracing the full cycle of knowledge" (ency-) becomes "instruction by and of chaos" (chao-) (75).[17]

Many iterations of the dark yet essential center guide the reader through discussions on the problem of evil, the phenomenon of the *métis*, the spirit as the third person of the Trinity, the universality of thought found in works from Pascal to Musil, on temperament and reserve (as both temperateness and *pudeur*), on meekness as a middle-virtue, on temperate and meek knowledge which has no conception of vengeance or reparations, on rationality that needs to be forgetful, on forgetfulness as a function of time. Among these allegories, two tropes in the section "Eduquer" stand out as striking examples of what I mean by a topological reading of the Serresean text. The first is the passage on "prescription," a legal term referring equally to the concepts of "squatter's right," statute of limitations, and the forgiving of a debt. Reflecting on prescription, Serres describes what cannot be inscribed *into* the Law as such, because it is prior to any historical writ or time-dependent law (212–17). He makes a distinction between two times. The first is historical time instituted by the Law and is therefore a cultural phenomenon. It does not forget, it is the time of revenge, restorations, reparations, and the eternal return. It is a stable invariant which we call "history or civilization." The domain outside the Law is what we commonly call "nature or natural law." Nature does not respond to the Law, but it, too, responds to the invariance of time. These two poles, Law and Nature, right and absence of right, become the foci of the "elliptical" prescription. This polarity will later serve as a basis for discussion in *The Natural Contract*. For now, prescription instates or authorizes time: "Prescription testifies to the essential action of time" (213).[18] Conceptually outside the law, not written into the codices of jurisprudence, preceding any legal status, it is, by the same token, also *an effect* of time. The Law recognizes the *statute* of limitations, the *right* of the squatter, and the *forgiving* of debts, only on the grounds of *time elapsed*, precisely on its own basis of time. Yet it can never inscribe into its codices the prescription, because the latter remains always prior to any "setting-down-in-script." What Serres explains here at some length with the help of the prefix pre-, is valid for the legal term of prescriptive acquisitions in both English and French.[19] The process of prescription is a "middle-position" (214),[20] because it brings into the Law, that is, into our historical consciousness, the notion of a time that is not passively periodic like historical time or nature's reversible time. It actively introduces forgetting: "Time passes and does not simply flow by passively; on the contrary, it forgets or effaces the deeds and rights" of any precursors (property owner, creditor, or prosecutor) (213).[21]

Prescription as a concept demonstrates the interaction between the local and the global: time introduces a minute, local difference that eventually checks global law. The entire discussion of prescription is in fact a demonstration of the irreversibility of time beyond the biological-social sphere, and of the interaction between the local and the global. One of the key notions of dynamical systems theory, the concept of sensitive dependence on initial conditions, states that small initial causes can have large, exponentially diverging effects. This is the tool with which we can explain the epistemological significance Serres attributes to a common occurrence in the legal domain: a single act committed at some point outside the law changes with time the given legal set-up—drastically and irreversibly. The consequences incurred *over time* far outweigh the significance of the initial act.

Modeled in this way, the legal concept of prescription becomes a metaphor in Serres's discourse for the continuing tracing or fleshing out of the instructed-middle. At the same time, it implies a third notion of time which is neither the classical reversibility of science nor the eternal return and stable invariance of historical time. This "new" time is creatively irreversible and, in one of its manifestations as prescription, changes the way we must begin to understand time both of nature and of culture. Prescription is Serres's linguistic material with which he molds a dynamical "middle-time," time proper for the tiers-instruit.

The demonstration is local, it concerns a legal (contractual) aspect of rights born of a natural dynamical process. But the demonstration changes imperceptibly to a global pattern when it passes to the central aspect of forgetting, or effacing the past, which characterizes prescription: "One does not forget forgetfulness which is an act [or writ] that is somehow unforgettable" (216).[22] We must add here that Serres uses the word "acte" which can also mean "writ" in jurisprudence; the English transposition unfortunately obliterates the elliptical reading of "acte en quelque sorte inoubliable." Finally, Serres underscores the global dimensions of prescription with a string of iterative presentations on the notions of reserve (179–83), meekness (185) and forgetting (212). This iteration guides the discussion to prescription as a middle-time which is the constitutive time of new ethical dimensions for our epistēmē.

As a metaphoric instance in the description of the education of the tiers-instruit, prescription is important because of the role time plays in its dynamical logic. Dynamicists in many fields have in common a new understanding of time and make it the center of their reflections. Peter Coveney and Roger Highfield, for example, give a survey of the notion of time in *The Arrow of Time*, a detailed account of its evolution in scientific and epistemological thinking. A very different approach, based on astrophysics and cosmology, is taken by Stephen Hawking in *A Brief History of Time*. Ilya Prigogine devotes his study, *From Being to Becoming*, to time as "the forgotten dimension" (xi) and speaks of its (re)discovery in connection with irreversibility as a period of scientific revolution "not unlike the birth of the scientific approach in ancient Greece or of its renaissance in the time of Galileo" (Prigogine, xiii). What he calls a "second or new time" vastly different from the reversible one of classical mechanics, "is no longer a simple parameter, as in classical or quantum mechanics; rather it is an operator, like those describing quantities in quantum mechanics" (Prigogine, xvi).

For the phase-space portrait of a nonlinear dynamical system, time is no longer one of the parameters that make up the state or configuration of the system. Rather, it plays an operating role in the behavior or transformation of the system. We may visualize time in its phase-space portrait as an operator acting on the system; its operation produces the phase-space portrait. Prigogine speaks therefore about the "timing of space."[23] To understand how closely Serres's text relates to these new scientific paradigms, it suffices to look at prescription as a "tiers temps" or middle-time, from the perspective of time as an operator:

> From the point of view of positivist laws or rights, our acts plunge into time [i. e. history as a social contract]; but from the perspective of prescription, they are born of, and formed by time, their true raw material.[24]

The analogy is quite clear: prescription serves as a Serresean demonstration of a new time that *operates* on the system of laws to make it less rigid and more dissipative, truly creative of evolving forms of human contact.

But prescription acts also as an operator on the complex system called "Le Tiers-Instruit"; for, when iterated "back" into its discourse, its meaning as a middle-time rapidly diverges into textual passages that the uninitiated reader would at first call "random." This randomness is inherent and therefore must be looked at, if the road to understanding is not to be abandoned. An example from the theory of complexity will again be helpful. Mitchell Feigenbaum studied the graphs of complex functions and their random behavior when they undergo iteration. An iteration scheme is basically a process by which a value of a given function, arbitrarily chosen as an output number from within the confines of the transformation rules governing the function, becomes an input number for generating the next value of the function. In "Universal Behavior in Nonlinear Systems," a highly technical article that offers nonetheless accessible nuggets of extraordinary insight for the nonscientific reader, Feigenbaum explains the functional iteration as a scheme by which "a certain function is reevaluated successively to produce a sequence of such numbers" or values (Feigenbaum, 5). Isotropically, Feigenbaum's iteration is the same as the mechanism by which Serres binds his textual tropes. The hyphen in tiers-instruit could thus be considered an intensely crunched formalism of this iterative binding. To Feigenbaum it is "obvious that no explicit formula will account for the vastly rich behavior [of a graph that emerges from functional iteration]" (Feigenbaum, 6). In the case of Serres's text, no explicit logic (or formula) will account for the vastly rich behavior of his textual iteration. Instead, a dynamical reading is needed to follow what, in *Les cinq sens*, Serres calls "logic by implication," analogous to the baker's transformations that such illustrations of complexity as the Cantor set or Smale's horseshoe express (Gleick, 51).

The iteration of the insight given by the Serresean prescription leads to further output values or random texts. A paragraph on the prescripted significance of epigraphs (216) produces, through input of "scription," a further random text evaluating the Gospel parable of the good Samaritan. This evaluation arrives at the question of authorship which, for its part, becomes the input value for a trinitarian passage that constitutes the end of the textual orbit of the middle-instructed: "I-Night" (217–34), "You-Day" (235–43), and "The Third Person: Fire" (243–49).

Studying the vastly rich behavior of a particular equation he chose as a case in point, Feigenbaum explains that

> while the iteration scheme [of his equation] is trivial [= simple] to specify, its nth iterate as a function of (x_o,y_o) [= the initial coordinates] is unavailable. Put differently, applying the simplest of *nonlinear* iteration schemes to itself sufficiently many times can create vastly complex behavior. Yet, precisely because the same operation is reapplied, it is conceivable that only a select few self-consistent patterns might emerge where the consistency is determined by the key notion of iteration and *not* by the particular function performing the iterates. These self-consistent patterns do occur in the limit of infinite period doubling and in a well-defined intricate organization that can be determined *a priori* amidst the immense complexity depicted [in the graph of the equation]. (Feigenbaum, 6)

Not only is Feigenbaum's mathematical description of the behavior of a system over time isotropically applicable to the schemata of *Le Tiers-Instruit*, but his rhetorical shifts can also help to elucidate the process at work in Serres's text: the "vastly complex behavior" of the tattered, fractal Harlequin leads to an "unavailable" nth information about who he really is and what his message actually contains. "Self-consistent patterns might emerge" if we let the parade of tropes run by our reading mind. But beyond the epistemological meaning of each iterated trope (or substory), we must understand that the pattern consistency of the Serresean tiers-instruit is not determined by the original system. The latter has been defined by the three equation-tropes (the left-right-hander, the swimmer in midstream, the goalkeeper on guard), isotropic to Feigenbaum's "particular function performing the iterates." Rather, "the key notion of iteration," translated in our text into the key notion of time as an operator, determines the "self-consistent" or self-similar patterns of the tiers-instruit. But the self-consistent pattern, this fractal image of the instructed-middle, "occurs in the limit of infinite period doubling" visualized by Serres in the luminosity of Pierrot.

As crucial to Serres's argument as prescription, and evolving from it through several metaphorical iterations, the second trope I want to discuss is an iterate of the passage on authorship (217), which itself is an output of the good Samaritan story as a non-juridical example of prescription (216–17). The passage titled "I-Night" addresses the question of the author, in both local and global terms. Author is first and foremost the first person "I" to whom Serres does not hesitate to assign an image and a face, himself. Critics often reproach him for these personal aspects in his writings, thus extending the debate over the subjective-poetic form of his more recent texts which works to undermine any objective-scientific claim they may have. I maintain that it is a reasonable gesture of appropriation which combines (1) the *I* he knows best, in the manner of Montaigne; (2) the reader's acquaintance with this *I* as author; and (3) the xth iterate of the initial input that included himself as a left-handed child becoming a right-handed pupil.[25] The authorial *I* is discussed in terms of the subject which becomes "sub-jectus" with its syntactical connotation and rapidly moves to adjectives and their importance for Serresean thinking. As the *I* identifies "itself-himself" through a long string of adjectives that slips inevitably and unhesitatingly to a "we" (220), a dissipative system appears:

> I am in reality all those [individuals] that I find myself engaged with, in and through successive or juxtaposed relationships; . . . I am therefore a mixed body (*un corps mêlé*), . . . like Harlequin's coat stitched together of adjectives, by which I mean terms placed side by side.[26]

The *I* becomes the topological configuration of the "viscous fluidity" (222) of never ending adjectives.[27] Clearly the argument has moved from a local singularity to a global level, or more precisely, to a global *pattern*, since a dynamical "coalescing" is favored over a stabilizing coalescence, as a model for this event.

In a very perceptive passage that parallels Serres's story of the "I" but is argued from the vantage point of recent philosophies of the fragmented or absent self, Alexander J. Argyros rejects the postmodern "demise of the self" without, however, "suggesting that the self can be equated with a Cartesian pinpoint. . . . [It] cannot be either a fixed essence or an isolated monad. It must be a dynamical system open to the world and especially to other selves" (Argyros, 290–91). Argyros goes further and suggests a dynamical model:

> If instead of the choice between . . . a transcendentally secure self and a disjointed schizo-self out for a walk, we were to picture the self as a chaotic attractor, then I believe that a model for selfhood might emerge that fulfills some of the correct requirements of classical conceptions of the self as well as the important contributions to the philosophy of self made by deconstruction and psychoanalysis. The self as a chaotic feedback loop locking-in on its attractor would be at once open to the world and to other selves, yet possessed of a deeply felt identity. A chaotic self would have a perdurable shape, the attractor, yet such a self would never repeat itself." (Argyros, 291)

The question concerning *Le Tiers-Instruit* is whether the Serresean *I* adheres to such a model.

Serres arrives at the harlequinesque "I" not so much by logical argument and deduction as by strings of tropological points (adjectives, prepositions, memories, geographical spaces visited, occupational possibilities, similarities with other animate realms, fluid phenomena [218–25]). His strategy is isomorphic to those of dynamical systems theorists who are forced to rely on numerical estimations and simulation (both made more powerful by computer graphics), because the appropriate mathematical theories are still lacking or in various stages of development. It is the absence of conventional logical tools in Serres's text that compels me to urge a reading isotropic to the phase-space portrait of a dynamical system. Each of the points that make up a Serresean string is like a point plotted in the textual signification space or meaning space of the system *corps-mêlé*, revealing the complete information on the system at that moment. There are many such strings, of adjectives, of prepositions, of tropes, of stories, and apologues; they function like various trajectories one can plot for a system's behavior depending on the arbitrary starting point of the plotting. But no matter how widely apart or separate the trajectories begin, eventually they all tend toward the attractor the system as a whole converges to. The dynamical system called "instructed-middle" is nonlinear by definition, since it is a deviation from the fundamental exclusion of the excluded middle, posited by the linear principle of Identity. Therefore, all the textual strings converge toward but never quite arrive at the "I" that knows itself as monstrously multiple like Harlequin, unitary and multiple at once. Like the *centon*, the *I* is a unique center and innumerable parts, simultaneously. The *I* includes all, without exception or exclusion; Harlequin is pan-cultural, pan-historical, pan-natural (225).

Inevitably the relentless iteration of the comedic "I"'s function in the text leads to the question whether God is part of this panorama. Is the concept of the divine one of the points on the orbit in the mental phase-space portrait of the "tiers-instruit"? This question is equivalent to starting the evaluation of the system from yet another initial state, and the answer is isotropic to watching this new trajectory develop and settle down to the attractor. From the paradigm of uniqueness inherent in the monotheistic concept, Serres develops what one may summarize as divine complexity: the one becomes trinitarian, the three become multiple in all the attributes by which we say the divine. These, in turn, lead to reflections on the notion of belief as opposed to knowing (228–30). Typically, Serresean belief becomes a middle-expression parallel to all the other "tiers-x" of the text. The all-inclusive multiplicities of this veritable pantology are called a "divine symphony," a pantheistic Harlequin who subsumes everything (231). This is a powerful image that, when transposed isotropically onto the phase-space portrait of the epistemological system "Harlequin," begs the question of its attractor. It is reasonable to conclude that a comprehensive, divine system must be a closed system. Serres's descriptive strings are creative enough to incite the reader to add to them in endless succession, but nothing new or inventive would appear. We might therefore compare the pantheistic Harlequin to a periodic and stable attractor. Though they can be very complex, such attractors are predictable.

Consequently, a pantheistic Harlequin would fail to adequately represent (imply) the middle-instructed and its fuzzy logic. Rather, he would fit the image of oppressive globalization that N. K. Hayles warns against in the course of her discussion on the local/global dichotomy of postmodern epistemological concepts (*Chaos Bound*, 233). Transposed into the cultural and epistemological domain, the closed system that appears to be emerging—God is all, all is God—is equivalent to the result of any deterministic approach or theory governed by rationalistic categories.

The symphonic drama staged by such a pantheistic Harlequin who usurps all forms, colors, and other categories appears to leave no room for unpredictable moments or *invention*. It looks therefore intriguingly similar to the "totalizing determinism" of the Serresean "static" statue, were is not for an nth iteration that reveals an *unknown* region toward which the dynamical harlequinesque system is attracted. There is a moment in the text when the reader witnesses the discursive staging of the imperceptible change from Harlequin into the luminous, fluid Pierrot, from divine closure to a laic position of open-ended uncertainties. Introduced by a double set of suspension points (rare in Serres's texts), this nth iterate opens onto the narrative "I" whose multiple attributes, all gathered at the nth point of the system's trajectory, suddenly reveal to the reader traces of a region of unpredictability. It is useful to cite Serres verbatim at this point:

> . . . but above all, I am sure and absolutely certain beyond any hope, that there exists a hole like an odd bit of straw in this massive and dense pantheism, a strange exception, the source of all pain; namely, that in this divine concert full of arch-noise, I and *I alone* am not God; this near-nothingness alone is not God, which gives new and poignant meaning to the old word atheist. [emphasis is Serres's] Here: no God. Only here, God is absent. My share of destiny is this place of atheism.—All is God, except he who writes this (*Tout est Dieu sauf celui qui l'écrit*) and then tearfully puts down his pen.[28]

The reader may have recognized the strange attractor, at least since the iterate of authorship (217). He does not receive locally ascertainable knowledge about the *I*, instead he follows an *I*-trajectory tracing out the strange attractor; the reader traces the *I* by following the author writing through his tears. The subject—Harlequin as the instructed-middle, Serres the author, or any *I* modeled as a dissipative system—is a system of knowledge or instruction tending toward a strange attractor. As such, he is the figure of "collective immortality," that is, the projection of infinite time within our epistēmē's finite destiny. This destiny, here named the "divine concert of all imaginable multiplicities," converges on the inascertainable *I*. The *I* attracts all knowledge, but itself is not part of it, yet is present in the process of its writing. This means that it is present in its determining "each subsequent state of the system presence," which is Weissert's description of the strange attractor as an aporic phenomenon (see p. 132). We will never know whether the tears are shed for joy at arriving at the middle-ground or out of desperation over the undecidable chasm that is *I*. The Serresean self is indeed the strange attractor for the instructed-middle.

If the *I* is traced by the writing subject Serres, then the attraction toward or settling down to this region of probable future states of the system is the writing of subsequent states of knowledge, in his texts. Since the *I* is now posited as a strange attractor, the subsequent states of knowledge of the instructed-middle become problematic from the standpoint of acquisition of knowledge, because by definition they are unpredictable. For Serres, however, this problem becomes salvation, because in the region of the strange attractor lies "new" knowledge: "My job is to write and say, not what I know, the boring, dead and spent [bits of information], but, on the contrary, what I do not know and what will come over me like a surprise . . . the unforeseeable which artists praise, the unexpected and, implacably, the improbable" (162).[29] The task of the middle-instructed is found in this region of global knowledge. It is the region of the strange attractor for a given dissipative dynamical system. Serres understands it as the region of *invention*.

Modeled in the fashion of the dissipative *I*, the concept of globalization no longer carries the threat of totalization or oppressive political structures (Hayles, 209–10), because globalization is revealed as an open system and inherently prevents totalizing closure. *Le Tiers-Instruit* tells the story of a human phase-space where each trajectory of the complex system called "divine concert" gives local information as it wanders through its ergodic space (of the text) and tends toward the attractor, but is "merely a trace of the attractor" (Weissert) and prevents systemic closure. Invention is Serres's name for this dissipative system with a topology so complex that, "unless the initial conditions are known to *infinite precision*, all known knowledge is eroded rapidly to future ignorance" (Feigenbaum, 21). Serresean invention founded on surprise, on an unpredictable and unexpected something, is the *I*-hole that the textual phase-space portrait of *Le Tiers-Instruit* plots along a delicate balancing act between the local and the global, in view of a new pedagogical ethics of the middle-instructed.

Serres ends his text with a touching testimony to another French figure who may be regarded as the modern antipode to the seventeenth-century Harlequin: Tintin and his creator Hergé (a pseudonym composed in reverse of the first two letters of his name, Georges Remi). Hergé's comic-strip figure enjoys universal appeal among the French and once again gives *Le Tiers-Instruit* a distinctive French flavor. The text that began with great emphasis on the harlequinesque multiple closes with the pierrot-like luminosity and transparency of an undefinable and unidentifiable hole that is *I* passing to *You*, the luminous friend Hergé without a real name, and then glides to the "Third Person: Fire," the region of blinding and luminous invention where the instructed-third comes into being.

Between Harlequin and Hergé-Pierrot, the baker's logic that is the grid for Serres's text works the multiple strata of cultural, scientific, and historical data and topologically stretches and folds them in order to suggest (*impliquer*) new dynamical solutions to age-old problems. In the process, Serres's rhetoric becomes a model for chaotic writing, while it is used at the same time for instructional purposes to describe and evaluate chaotic systems. Borrowed from a traditional French theatrical theme, and elucidated for modern consumption through an encoded application of complex dynamics, the Harlequin story demonstrates that an orderly system can successfully transfer its knowledge to the next generation only when it is willing to accept multiple dissipative learning processes, without reducing them to its own binary mold. It outlines an instructional nonlinear system with non-closure as a valid form of education. It brings the excluded middle into play without any obligation to choose between exclusion or inclusion, a development Serres calls a "secular miracle." We must understand this expression once more as an epistemologically new concept that plays on current fundamental debates in France, but goes far beyond the question of secular public governance vis-à-vis religious laws (see chapter 3 of the present study). The "bound" chaos of tolerance levels the insurmountable walls between the sacred and the profane-secular. Serres pleads for this type of instruction, as a complementary element to traditional linear Euclidean learning and as the only guarantee for the most precious freedoms of all—the freedom of inventive thought. Closed, repetitive systems smother invention, dissipative systems nourish it with their self-organizing tendencies.

Le Tiers-Instruit is a blue print that suggests for a new educational philosophy on a global scale what scientists, in particular dynamicists, have suggested for purely pedagogical reasons. Robert May, for example, makes "an evangelical plea for the introduction of [higher order] difference equations into elementary mathematics courses, so that students' intuition may be enriched by seeing the wild things that simple nonlinear equations can do . . . the bizarre behaviour exhibited by the simplest of discrete nonlinear systems" (May, 459, 567). The plea for a methodology that encourages the expression of nonlinear intuition in mathematics instruction is isomorphic to Serres's call for the inventive thought of chaotic instruction, and is the more intriguing for its quasi-religious rhetoric (evangelical plea) on the part of the scientist, since Serres hopes for a secular miracle. Both speak of the enrichment of their students' *intuition* or *invention* when confronted with chaos. The poetic irony of Serres's argument for dissipative, dynamical learning lies in the fact that, after *Statues* where death plays a crucial role in the dissipative process, Serres should now choose a clown and practical joker to portray the ideal dissipative teacher.

6

Time and Earth: Reading *Le Contrat naturel*

Alert and healthy natures remember that the sun rose clear.
—Henry David Thoreau, *Walden*

Seen from the perspective of Michel Serres's global work, *The Natural Contract* (1990) is a text that invites a slightly different approach. It has no immediately apparent connection with literature or science per se. Its title borrows from Jean-Jacques Rousseau's *The Social Contract*, although it is definitely not a sociopolitical tract. But it has everything to do with jurisprudence, legal justice, the philosophy of power, and the question of right. It argues in favor of a broadly outlined ecological position without explicitly referring to current ecological movements or their political reverberations, particularly in Europe. The absence of an outside objective observer who would hold this ecological position gives it its specific character. Its rhetorical style, however, remains true to Serres's writing in the genuinely French tradition: it is a scientific and philosophical treatise executed in the inimitable and elegant literary essay form that has become the philosopher's hallmark.

In *The Natural Contract*, Serres lays the philosophical foundation for what I would characterize as a new ecological concept. In this context, "concept" is used in its original etymological structure of *con-cipere*, meaning to grasp or catch together, in the concrete fashion implied by the composition of the book's title. In order to explicate what its elements are, I begin with the opposite notion of a truly ecological concept. In a remarkable article which appeared in *Time* magazine (June 17, 1991), the syndicated columnist and essayist Charles Krauthammer argues for a conservative environmental policy that would preserve nature to the extent that it is of use to the human species. He concludes his essay stating: "Nature is our ward. It is not our master. . . . It is man's world . . . nature will have to accommodate. Man should accommodate only when his fate and that of nature are inextricably bound up" (Krauthammer, 82).

Reflecting a hierarchical thinking, these remarks go to the heart of the division between the anthropocentric view and Serres's ecological concept. The limiting conditional "man should accommodate only when . . ." is a logical fallacy, from the Serresean viewpoint, which prevents us from making a now critically and morally necessary, radical reassessment of the culture-nature dichotomy that informs all human epistemological endeavor. Instead of asserting that "man should accommodate only . . . ," the Serresean ecological concept implies that humankind's postmodern fate and that of postmodern nature are, *from here on forward*, inextricably bound or grasped together into one fate. A close reading of Serres's text shows that the term "postmodern"—whatever definition one dare give it—applies in fact to nature as well. It is the phenomenon of a planet physically and biospherically altered by human intelligent intervention *on a global scale*. Henri Poincaré's dictum on human thought as merely a "flash of lightning" in the midst of a long and vast night of cosmic nothingness, may no longer hold for the future, even if Poincaré follows up with the reassurance "that this lightning is, however, everything" (as quoted by Prigogine and Stengers in the preface to the second edition of *La nouvelle alliance*, 26–27). This "thought-flash" is not only all-important in human terms, but equally determining for the planet Earth, perhaps the only heavenly body in the vast cosmos to be permanently altered as much by cosmic-mechanistic events as by the products of a flash of thought and free will. In this sense, earthly nature is forever marked as different from all other cosmic bodies. For humans, this difference can be conceptualized as postmodern nature, a singular cosmic state entirely restricted to our planet.

The question addressed by *The Natural Contract* brings into focus the future relationship between humankind and Earth, modeled on historical contracts between human and human. These models are familiar to us because culture as opposed to nature is entirely composed of contractual building blocks. In addition, the arguments of this text gravitate toward Jean-Jacques Rousseau's writings on social relations among humans, as around an invisible pole. It is somewhat ironic that a compatriot of Serres, Luc Ferry, neglects to account for the French sources from which Serres primarily draws his authorial strength and power of persuasion. In his study *Le nouvel ordre écologique* (1992), Ferry summarily relegates Serres's propositions to the camp of those who advocate the judicial rights of nature and regard the earth as a legal agent and a "juridical alter ego" of the human being (Ferry, 257).[1] More specifically and noting that Serres teaches in California, he states that Serres's theses comprise "an authentic crusade *in the American mode* against anthropocentrism in the name of the rights of Nature" [emphasis added] [une authentique croisade à l'américaine . . . contre l'anthropocentrisme au nom des droits de la nature] (Ferry, 32). Ferry is referring to alternative ideologies that dominate various ecological movements in Europe and North America. But he is particularly critical of the American brand of ecological zealotry which readily transfers basic concepts of the Anglo-Saxon judicial system, which relates the protection of rights to that of interests, to the biosphere, so that its interests—the well-being of fauna and flora—metaphorize rapidly and without further reflection as to political consequences into rights of animals and trees as conceptual entities equal in all respects to the rights of the human being (Ferry, 92). For an intellectual steeped in a political and sociological tradition based on the "Déclaration des droits de l'homme," which has guided public thinking in post-revolutionary France, the implied mutations of humanist ideals are ominous. Ferry argues in effect that proponents of such an ecological fundamentalism are dangerously close to an ecological dictatorship and weaken the chances for a democratic humanism, which is the only acceptable underpinning for a sane and "democratic ecology" (*une écologie démocratique*) (Ferry, 241). This democratic ideal is part and parcel of "a non-metaphysical humanism" which is inalterably tied to the historical emergence of political laicism. The question of ecological responsibility is therefore intimately related to the construction of a "laic and democratic world" (*l'univers laïc et démocratique*) (Ferry, 246–252). Ferry's ecological philosophy is in this sense very close to Serres's epistemological critique of Western thinking. Both rely on the thoroughly French notion of *laïcisme* and both enlarge the concept so that it becomes applicable to a new way of conceptualizing the human and the world. It is therefore regrettable that Luc Ferry does not take into account Serres's use of the laic perspective to recover the excluded "other" and undo the hegemony of all sacrosanct ideologies and inner sanctums. It points also to the necessity to read Serres globally as the present study tries to do, so that the understanding of new concepts such as the laic perspective carefully derived and presented in *Les cinq sens* would further the global meaning of a text like *The Natural Contract* where laicism is not explicitly discussed.

Although many of Serres's premises echo those of American ecological groups whom Ferry accuses of radically discrediting Western humanist tradition, the critic does grant his fellow philosopher Serres the right to poetic license in the interest of persuading a lethargic public (Ferry, 245). And he readily concedes that Serres's "literary metaphor" of a natural contract is related to the "famous social contract of eighteenth-century philosophers" [analogue au fameux contrat social des philosophes du XVIIIe siècle] Ferry, 150). But Serres's debt to Rousseau goes much deeper.

Rousseau's dramatic opening statement of part 2 of the *Discours sur l'origine, et les fondements de l'inégalité parmi les hommes* posits the recognition, by *others*, that an individual has the *exclusive* right to a local "here": "The first person who, having fenced off a plot of ground, took it into his head to say this is mine and found people simple enough to believe him, was the true founder of civil society" (*The First and Second Discourses* 141).[2] The acquisition of an exclusive right is based on the act of taking possession. But it is finalized and legitimized only when two parties enter an agreement, the individual claiming possession, and the "others" approving the claim. The contract thus drawn up is the constituting element for civilization. One must understand the full impact of this *contractual* activity on the historical thinking that motivates the author of *The Natural Contract*. For not only does it determine forever onward the relationship of contractual partners, but it also starkly recalls the *exclusionary gesture* that any contractual activity produces. The exclusion is not a mere, albeit necessary by-product, but the sine qua non of any binary understanding: This is mine; nobody else can enter the "here," this terrain which remains here for me as owner to hold and for you as my contractual partner to respect as mine. Any third-comer is excluded by dint of a suspending gesture; Rousseau posits the potential actions of the latter in a conditional mode or speculative mood, as he continues his apocryphal account of civilization's founding.

The necessity to exclude something or someone represents the core characteristic of our traditional contractual consciousness: The corollary to any mutual agreement is the denying of the very results of the agreement to any and all outside the contractual partnership. This means that the contractual conscious is always local or acts always locally. From the time it was established as the cornerstone for human collaboration, humankind's historical activities have always been local, or particular and limited achievements, no matter how far-reaching their effects. This local aspect of our cultural work is inherent in human nature and at work at the very origins of all cultural and symbolic activity, as René Girard described in great detail as part of his theory on the process of hominization of the human animal (*Violence and the Sacred*).

In the Girardian elaboration of the origins of symbolic-religious activities as seminal manifestations of the hominization process, the excluded entity is first and foremost a human, the original victim, set apart from all others participating in a mutual agreement. The human victim is later replaced by animal sacrifices or other scapegoat emissaries as acceptable substitutes for what was originally the singular culpability of one human being. His expulsion and murder constituted a guarantee, *against* mutual violence and potential mass extinction, and *for* the efficacy of the original, unifying pact between all members of the human group.

Serres's ecological concept does not contradict the theory of the scapegoat mechanism, but enlarges the implications which Girard pursues for the foundation of religious and cultural systems. As presented in *The Natural Contract*, Serres's thesis is the elaboration of a different perspective concerning the scapegoat mechanism. It neither focuses on the relationship between sacrificial victim and social order, nor on its contractual nature evident in the process of sacralization and eventual divinization of the victim as the guarantor of social order and peace. Instead, Serres steps out of the dynamics of this process in order to focus on the expulsed "other" of the contract between emissary victim and society; that is, he brings into focus the excluded "outside," which does not participate in the hominization process. This "outside" becomes logically an "excluded middle," (*le tiers exclu*).

For Serres, the primary and globally excluded entity, given the fact of human historical existence, is and still remains the planet Earth. Since she (*la terre*, *la planète*) is global, she never could be recuperated into our epistemological stomping grounds. Cultural contractual consciousness is exclusively based on local, cause-specific considerations. As Serres writes at one point: "Human history could take pleasure in itself (*jouissait de soi*) in an a-cosmism of inanimate matter and of other living things" (39). The use of the French term *jouissait de soi* has not only onanistic overtones and subtly underscores humankind's reckless abandon that forgets the rest of what is "out there," but it also reflects Serres's urgent appeal for a *new* concept in contractual thinking that would wed Earth and humankind in a partnership marriage. The new relationship would no longer be one of warden and ward, such as anthropocentric ecologists would have it. In the last pages of *The Natural Contract*, Serres relates his experience of the aftermath of the 1989 California earthquake (114–15), in words that leave the reader with a vague feeling of being a voyeur to a passionate encounter between the trembling Earth and humankind metonymically presented by the author himself. This passage no doubt contributed to Luc Ferry's characterization of *The Natural Contract* as a "romantic inspiration" (in his review of Serres's text; *L'Express* April 1990:61). One may think perhaps, along with other critics, that Serres is settling ever more into a lyric and poetic mode, somewhat parallel to religious or teleological tendencies many skeptical scientists attribute to the Gaia theory.[3] But *The Natural Contract* first lays the foundation for a rigorous scientific-judicial appeal to its postmodern reader. Only then does the vision of an as yet unrealized world created on the basis of a natural contract between the Planet Earth and humankind move the author to passionately poetic expressions, equally serious in their intent. One does injustice to this text, if the logical strength of either its scientific-historical aspects or its poetic power are ignored.

The notion of prescription plays a prominent part in the arguments that complete the discussion of *Le Tiers-Instruit*. Now however, it is isotropically moved from local and topical singularities of squatter's rights, limitation of statutes, and forgiving of debts, to the global level in a topological sense. In the Serresean application of its original meaning, prescription models a right based on a notion of time that is not connected to historical periodicity. In *The Natural Contract*, this right is globalized: subject and object, human and earth, enter a contract which falls outside all contractual modes, because it knows no exclusion. Yet it will recuperate and subsume all laws, that is, the Law. Prescription binds historical time codified in the Law to a dynamical, ahistorical time with which the Law as a social contract must form a partnership. The project of *The Natural Contract* is to present dynamical time in the most applicable and concrete form we are capable of understanding today, namely, as a natural contract, sealing a union between the human subject and the physical object. It puts into a proposal with practical applications for the future what Serres has gleaned from an ancient fabled setting called "Ulysses' circum-navigations," a nonlinear dynamical system that invents new knowledge at every turn, without closure. We call it "epic" and ignore the fact that what we so designate is the story of *a new time* (*Les cinq sens*, 290).

Two phases compose Serres's demonstration in *The Natural Contract*. The first concerns science's preoccupation with explanations of the world that appear to yield less and less true knowledge of reality as such, because they are obtained through a series of "local fractured systems and models of analysis" (Hayles, 2). In reaction to this process of diminishing returns, science usurps the Law or Right (*le droit*), whenever it wishes to make a newly attained local knowledge into a canon, or at least into part of the epistemological canon our cultural world consists of. Isabelle Stengers used the case of Galileo to argue a famous example of such usurpation. In this scenario, she sees the scientist no longer as a victim standing accused before the papal tribunal but as the unrecognized manipulator of the relationship between science and power, the very power the Church insisted it held absolutely through the Law (Stengers, 12–16). Serres even argues that scientific invention and/or discovery are preceded and often launched by the Law. In his familiar apologue style, he cites, among others, the annual reconfiguration of the Nile Valley into arable plots after each seasonal inundation, an activity that gave rise to geometry, according to the Greek historian Herodotus (51–53). We see the usefulness of this example only, if we recognize in the "harpédonaptes" the representatives of pharaonic law, acting as land surveyors. Pharaonic decree determined the measures to be used, which in turn created, over generations, a geometric order leading to geometric abstraction. Another example is the drama played out in the Garden of Eden, where the Law, in the form of the original divine injunction, "Thou shalt not eat . . . ," is father to the desire of knowing (57–58). Repeatedly, Serres describes science and Right in an adversarial relationship; he, too, recalls the case of Galileo versus the papal court, and the plight of Lavoisier before the French Revolutionary Tribunal which condemned him to death in 1794 (61–62; 81–86). Throughout human history, Serres maintains, judging reason (Right and Law) and demonstrating reason (science) play a game of one-upmanship, with victory claimed, now on one side, now on the other (65).

This fundamental tug-of-war changed drastically with the advent of the twentieth century. The steady quantitative increase of scientific knowledge since Galileo resulted in upending the balance of this dialectic game. Born of the Law, science today holds all Right, all titles; it reigns triumphant (84). Serres goes further, maintaining that "science takes the place of law and establishes its tribunals, whose judgments, henceforth, will make those of other authorities seem arbitrary" (86). For the postmodern era, the novel and *qualitatively different* aspect of this development lies in the critical juncture of the realm of knowledge of the objective world (the sciences) with the social contract (the Law) that is our cultural history. Serres speaks of this juncture in terms of a frontier, a watershed, a critical accumulation or saturation point, a global change (13).

For Serres, the conflation of *res* and *reus* (object and cause) is functionally analogous to "phase transitions." In physics this term refers to changes of matter from one state (solid, fluid, or gaseous) to another; what happens dynamically at the boundary between two states can be mathematically difficult to analyze, because the process tends to be nonlinear. Analogously, complex systems theorists understand phase transitions as points or thresholds where the behavior of non-linear systems hovers "between chaos and order." More significantly, they realize that at the "edge of chaos" where systems are poised in "the narrow transition region" between order and chaos and can go either way, the potential for information processing and manipulation is enormous. They see the edge of chaos as a "time" in a complex system's fate that is full of "immanent creativity," and have shown that such a system adapted and "honed the efficiency of its rules as it went" (Lewin, 51–55).

The notion of "edge of chaos" or phase transition is rich in its implications for other domains of knowledge. Complex systems theorists connect it with Prigogine and Stengers's concept of systemic self-organization. Stuart Kauffman uses it to discuss biological evolution and formulates a working hypothesis that "life exists at the edge of chaos" (Kauffman, 26). He speculates that this fractal basin boundary, "this compromise between order and surprise," may also be a useful theme for the formulation of a "general law" with possibilities for a "deep new understanding of the logic of democracy" (Kauffman, 27–28). Serres's argument concerning the changing relationship between object and cause with its potentially rich and creative promises is therefore a most critical phase in the development of his concept in *The Natural Contract*. The problematic equilibrium between science and the Law, which synecdochically stands for the relationship between nature and culture, or between object and subject, has shifted *from the local to the global level*: today, the human contractual conscious spans the globe; no longer can we locally exclude, when we enter a given contract with a new social partner; we have contracted all possible partners. Divisions into third, second, and first worlds notwithstanding, culture has become absolute and global today, as it egotistically (anthropocentrically) lays a grid about the entire earth, to the total forgetfulness of nature. At the same time and in its efficient pursuit of the object, science has mastered the global object, the planet Earth. Important unanswered questions and unsolved problems in genetics, nuclear physics, astrophysics, and so forth, notwithstanding, from the epistemological standpoint that is Serres's, science has become global. It does alter the Earth itself, and can do so even to her very extinction. As a global community, we have entered a phase of transition, we live a critical and chaotic era.

Seen from this vantage point, the conflation of *res* and *reus*, object and cause, nature and culture, is the clashing of two Titans: two globalities (*globalités*), having ignored each other in their preoccupation with local processes, are suddenly thrust together, at the risk of cultural extinction if we continue to ignore nature; or at a high risk for the Earth as we know her (still). In either case, humankind is the big loser.

The second phase of Serres's argument begins with the idea that our contractual consciousness has treated the earth as an excluded middle or "third" (*le tiers exclu*) which in Serresean epistemological context is a parasite. The global voice of our planet is drowned out every time contractual discussions between two potential partners occur. Once upon a time, the Earth's concerns were perceived as voices and signs from the gods, as beneficial or malevolent divine forces inhabiting both animate and inanimate nature and regulating all life on Earth, including human existence. Animistic religious forces were gradually transformed into mythological deities and lost their magical powers when these, in turn, metaphorized into transcendent divinities. From the perspective of Serres's ecological concept, on the other hand, projecting divine forces into nature and later into a transcendental realm was a conceptual means of excluding natural phenomena from human life's contractual existence.

Human experience having been neatly divided into two spheres, the immanent and the transcendental, the planet's concerns became, once again, annoying background noise (hurricanes, earth quakes, droughts, etc.) against which modern science harnesses all available knowledge in order to neutralize its disturbing effects. Since Newton, science has replaced the spirits of woods, streams, and springs and presents the empirical world to us through abstract models, so that this nuisance may be less intruding. All the while, we have overlooked our own parasitic existence that feeds on the earth. As our contractual culture changes from a multitude of local forces to a single global force, our parasitic activities become threateningly destructive and forever more deadly for the host. As the annoying background noise increases correspondingly, by way of deadly, global backlash (ozone depletion, climatic changes, extinction of species, global warming, etc.), there is one and only one option left: Serres suggests that the two globalities must enter into a symbiotic relationship with one another. The two must stop acting parasitically and become mutual symbiotes. Having become equipotent with nature's global forces, global culture must seek to enter a contractual partnership with the Earth, because there are no longer any potential partners left.

In this context, Earth resembles Gaia, the earth-model of Gaian scientists. One of the harshest criticisms leveled against the Gaia theory deals with the tendency by some of its adherents to view the Earth as a system capable of optimizing intentions, or even as a pseudo-divine entity. It is interesting to note that Serres's ecological concept skirts these ideas, but in an epigenetic sense only. Three years earlier in the initial chapter of *Statues*, Serres, the historian of science, had formulated an analogy for the archaic sources of the sciences. The shocking juxtaposition, of the 1986 Challenger disaster of the American space program and the ancient Carthaginian sacrifices of children to the idol Baal, served to snatch the reader from a long and hallowed tradition of the "noble" and "rational" sciences, and instead to confront her with the scandalous fact that the most advanced sciences and technologies of today still harbor mythic and anthropological traits (*Statues*, 47).

There is, of course, an entire literature on the interpenetration of myth, religion, and early science. One of the more recent and daring contributions to this field can be found among the themes discussed in *Trialogues at the Edge of the West: Chaos, Creativity, and the Resacralization of the World*, where one of the trialogues' participants states: "If we examine the history of early modern science, we discover that some of the major movers and shakers were being guided and directed in the formulation of early science by discarnate entities" (Abraham et al., 95). Among several names, the authors cite Descartes and Newton. Similarly, the author of *The Natural Contract* reaches far back into the mythological and magical past out of which science sprang and which modern science, true to its schooling in the great Enlightenment period, has only too eagerly forgotten and too scornfully disowned. Where nature was once perceived as a patchwork ensemble of local deities, demi-urges, and spirits, Serres now recognizes a *global* figure having evolved from a medley of magico-mythic entities into a concrete, empirical reality. He implicitly fathoms the (pre)historical past dominated by magical world views and is able to extract from it what he calls elsewhere "a global object: Earth" (*Conversations*, 116).

For the purpose of the argument put forth in *The Natural Contract*, the model for the global object is Earth as a contractual partner. The two "globalities," nature and culture, must act contractually, that is, like two localities (*localités*) but without excluding, which means without "acting" local. This paradox makes it imperative to conceptualize something other than the binary concept local/global or nature/culture. A third term is introduced, precisely the *natural contract*. A contradiction for traditional thinking, we have to learn to manage this paradox, both culturally and scientifically, in order to cancel out the mutual exclusion that has always marked the relationship between the natural and the cultural realms. The human share in this effort consists of the reintroduction of the "excluded" middle (*le tiers exclu*), but not as an included entity, for this would imply a dialectic flip-flop leading to yet a new exclusion. Rather, Serres conceives once again of a "tiers-instruit," the concept by which any exclusionary gesture is suspended in an emergent global phenomenon. It is nothing less than the call for "a new human" (94–96).

Rousseau's dramatic genesis-story of the original contract, agreed upon by a "first" individual and those recognizing his claim, also offers a glimpse of the new human. The Rousseauistic scene of the very first claim to private property is followed by a second one in which the protagonist is an "excluded" middle. This third person does not participate in the contract between property owner and those acknowledging the ownership. He is a human scapegoat whose guilt rests squarely in the fact that his action is not taking place in reality: "What crimes, wars, murders, what miseries and horrors would the human race have been spared by *someone who*, uprooting the stakes or filling the ditch, had shouted to his fellow-men: Beware of listening to this impostor: you are lost if you forget that the fruits belong to all and the earth to no one!" [emphasis added] (*The First and Second Discourses*, 141–42)[4]

The author of *The Social Contract* had a clear vision of the danger inherent in the exclusionary gesture of the human contractual conscious. But for the hypothetical third actant's protests, humankind would have been spared the vicious cycle of violence and exclusion of historic proportions. Nonetheless, Rousseau himself practiced the exclusion with a stylistic ploy: The "missing" action of the third actor is purely speculative and expressed by a conditional mode with a strong implication of nonrealization, while the actions of the contractual partners are expressed in the historical past (the *passé* simple). In the discursive opposition of these two modes, the writer's refusal to conceive the third person's action on the level of historical reality is suspended into a supposition of an "if only he had been there . . ." The suspending gesture serves to attenuate the pure exclusion of the third, in a "conditional" attempt to include the "excluded middle" in the historical fable of a foundation, however tangentially. In the context of Serres's discourse, Rousseau's illusory third functions as a projection for the "instructed third" whose essence will prevent exclusion per se and whose wisdom will bridge the paradox of the natural contract. Rousseau's ever so cautious rattling of the cage of stable reality foreshadows the nonlinear complexity of postmodern epistemological thinking. From this perspective, *Le Tiers-Instruit* tells the story of the descendant of Rousseau's suspended "third."

Following the outline of *The Natural Contract*, one can formulate three clauses for such a treaty: (1) As is the nature of both the "social contract" to which adheres the political collective and the scientific contract which unites the collective of experts, the natural contract is an unwritten pact. "Virtual and unsigned like the first two contracts, since it seems that the great fundamental contracts remain tacit, the natural contract recognizes and acknowledges an equilibrium between our current power and the forces of the world" (46). What is a virtue for Serres, namely, the intrinsic value of a forever-unwritten agreement, presents a weak point for Ferry for whom "this is more a metaphorical fable than a case of rigorous argumentation. It seems indeed difficult to attribute a proper meaning to the contract as proposed by Serres" [il s'agit là d'une fable métaphorique plus que d'une argumentation rigoureuse. Il semble bien difficile, en effet, de conférer un sens propre au contrat proposé par Serres] (Ferry, 151–52). As such the critic doubts that even "Serres would seriously consider that man and nature could ratify together an authentic 'contract' where they would act as equals; but that is what he appears to suggest nonetheless" (Ferry, 245).

But Serres is quite explicit on the meaning of the concept "contract." From its etymological root of "binding," "contract means that some collective is working together to pull or draw something" (*Conversations*, 88). He speaks of this collective pull as an "inextricable and transparent knot" (*Conversations*, 152) which links nature as represented by science, and culture expressed through the Law. In *The Natural Contract*, he dedicates the last chapter to the demonstration of the essential inextricability of the knotted pact, via a series of apologues. Serres exposes himself thereby to criticism that he is retreating into poetic persuasion, when a rigorous scientific defense may be needed instead.

An example of an identical argument that stylistically, however, adheres to moderate views of a more sober discourse on complexity is *Malicorne: Réflexions d'un observateur de la nature*, in which Hubert Reeves, French Canadian astrophysicist, who works with the French research group CNRS, expresses a similar conviction about the absolute necessity of a "legislation" for a certain organization between human and nature. But Reeves argues convincingly from the standpoint of complexity and its enormous importance for articulated theories on self-organization of dynamical systems and emergent phenomena. Reeves's text expresses the equivalent of the Serresean inextricable knot, but through the notion of a legislation commensurate with the "infinite complexity" of nature. Since human intelligence is the summit of this complexity, Reeves continues, "the jurisdiction for this organized complexity must adapt itself to the inextricable cluster of the human psyche" [Sa juridiction doit s'adapter à l'inextricable écheveau de la psyché humaine] (Reeves, 163). Stopping short of calling it a contract, the author argues that such an adjudicated complexity would "assure the co-existence, and simultaneous development of the individual, the group and the planet" [Il faut assurer la coexistence et l'épanouissement simultané *du moi, du groupe et de la planète*.] [emphasis is Reeves's] (Reeves, 163). By establishing a direct and concomitant relationship between human jurisprudence and complexity as a fundamental dimension of nature, Reeves implies, in effect, an equilibrium between present-day human power and the forces of nature (Reeves, 152). Assuring the stability of this equilibrium is also the driving force that motivates Serres to tell a fable of knots and cords with a moral call for a natural contract.

(2) "The natural contract acknowledges above all the new equality between the force of our global interventions and the globality of the world" (46). Expanding on the first "clause," Serres refines the notion of equilibrium of forces as a de facto *physical* reality, by consigning to it a *legal* quality: Since the two forces are balanced evenly in their commonly shared global character, they must also have a shared *equality*, that is, equal right must be acknowledged for both sides, if a clash of the two globalities is to be averted. The notion of Right accorded to the object is, of course, the controversial aspect of this exposition.

(3) The natural contract abides by the crucial principle of *rendering* (89–91): The Earth gives us the object, nature offers that which is given in all its complexity and myriad forms, without exception; modern technology permits us to engage in global taking. Humans, in turn, must globally give back, in a spirit of what only we humans are capable of formulating morally, because it is only right to do so. "What can we render to the nature that gives us birth and life?" asks Serres and then answers instantly: "The totality of our essence, reason itself" (90). This is the contractual basis for a "new call to globality" (93). It allows for two

globalities not merely to manage their global differences, but to stipulate *de jure* solutions to conflicts of interest with global consequences.

By tying together, in a truly symbiotic relationship, reason understood through its sociocultural manifestations, and nature as we know it through scientific inquiry, Serres places the human being in a partnership where proponents of anthropocentric ideologies misplace him as a master, and those of the Gaia theory forgot him altogether. Human *reasoning power* becomes contractually (rightfully) an agency whose services the Earth may lay claim to, and avail herself of. No longer is human life merely one entity among many of the biota that compose the Earth. The partnership with human reasoning assigns to Earth cultural and moral dimensions which the Earth may rely on so that justice be done to her. This is an ecological concept that daringly embraces teleological dimensions, because it places human reasoning power at the service of the earth's concerns. Serres proposes to contractually bind the multiple manifestations of that power to a historical or optimizing consciousness which organizes these concerns. Such a construct permits scientific propositions to remain rigorously scientific without teleological contamination. It legitimizes the Serresean ecological concept.

Equally circumspect and open to an ecological concept, Reeves's reflections in *Malicorne* parallel those of Serres but are argued from the standpoint of complexity. For Reeves there is no doubt that the human being is part of nature as a product of biological evolution; he is in nature. But because of his "extraordinary level of performance" (*son extraordinaire niveau de performance*), one can consider him equally outside nature. Among all the strategies of complexification, nature has come up with "the strategy of intelligence" (*la stratégie de l'intelligence*). It allows humans to disregard genetic injunctions with which nature programs its powerful but mindless creative impulses into a semblance of locally ordered systems (for example, bee and ant colonies). Reeves concludes that, since the human is part of nature but has been "evolutionized" out of nature and has surpassed her, "nature becomes conscious [through the human] of the predicament she has boxed herself into" [on voit . . . *la nature devenir consciente de l'impasse dans laquelle elle s'est engagée*" (emphasis is Reeves's)]. Nature has arrived at an explosion point of her creative fervor (Reeves, 156–62).

With reasoning power being the human contribution, Serres's stipulation of mutual rendering is superbly justified by Reeves's viewpoint: Human intelligence evolved out of nature through the process of complexification. The appearance of human intelligence as part in and of nature enables him to work outside nature, which in the Serresean model constitutes the ability to be a partner with nature. From the inverse perspective, this gives nature a reasoned voice, by which she becomes a partner of humankind. Reeves's formula is a particularly persuasive one: "Forced to surpass herself and to abandon this obsession with [creative] performances she had dedicated herself to, until now, *nature invests in the domain of values*. Through the arrival of a moral sense in humans, she opens her eyes and becomes responsible. The human is nature's conscience" [Elle (la nature) se sent forcée de se dépasser elle-même et de quitter cette obsession des performances auxquelles elle s'était jusque-là confinée. Elle investit le domaine des valeurs. Par l'avènement du sens moral chez les humains, elle ouvre les yeux et devient responsable. L'homme est la conscience de la nature] (emphasis added) [Reeves, 162]. Both, nature and human reason, will act in a self-regulating relationship that, if Serres's urgent appeal is heeded, is sustained contractually rather than systemically. It is therefore futile to read *The Natural Contract* in hopes of finding a definitive contribution to existing ecological movements, in the sense of ideological encouragement or concrete suggestions for solutions to environmental problems. Though "system" and "contract" share conceptually the same prefix alluding to "composition" or "assemblage," the idea of a contract assigns a legal frame to two systems that in the past structured themselves locally and globally respectively, yet today have both become global entities that need to *con*struct, or face global *de*struction.

Historians may see in the arguments of both Serres and Reeves the workings of an "anthropic principle," that is, "that the universe looks the way it does at least partly because we are here to do the looking, and because we are the way we are" (Brands, 507). Brands mentions this in passing while exhorting historians not to rely unconditionally on scientific paradigms, not even on the most promising ones in the realm of fractals and chaos theory, when attempting "to explain why people are the way they are" (Brands, 509). Serres and Reeves develop the anthropic principle in the course of their reflections, Reeves through the modern dynamical notion of "complexity" and Serres by recognizing the fundamental linkage between science and right, *res* and *reus*, object and subject. Historians should likewise not forget what modern natural science has been practicing: "Henceforth, physics re-examines what classical science denied in the name of the reversibility of elementary processes: the notions of structure, function and history" (Prigogine and Stengers 1986, 44). The Serresean ecological concept is a constant reminder that, in its full range, the anthropic principle includes the Earth. From their perspective, Prigogine and Stengers call this a dialogue:

> The task is not to reduce nature to a mute identity, but to construct the meaning of the kind of questions we can ask of her. As in all true dialogues, the crucial points are those where we can recognize and incorporate into our representation of the other [the object], what until then we had believed to be able to determine exclusively by our own subjectivity. (Prigogine and Stengers 1986, 26)

Whether it is understood as a contract or a "true dialogue," the issue of ecological stewardship is put in either case on a level that should be approachable for both scientist and humanist. Why this approach *must* occur is Serres's preoccupation in *The Natural Contract*, a text to which he refers repeatedly in *Conversations on Science, Culture, and Time*, where he calls it a global fable (88). In these conversations with Bruno Latour, he repeats the metaphor of the play of strings (*le jeu de cordes*) that dominates the last pages of *The Natural Contract* (107). It enhances the final demonstration of the constant informational exchange, from the local to the global and from the global to the local, as the primary dynamics for a natural contract.

What must finally be asked then, is a question of style: Why does Serres expose himself to criticism that his discourse lacks scientific rigor, and why does he favor instead what he calls "the everyday language" of "stories?" Latour gently prods him for an answer with the challenge that he, Serres himself, has created an extraordinary number of misunderstandings concerning his writings (*Conversations*, 72). To which Serres retorts that Plato, Leibniz, and many others abandoned their technical discourses when the going became too complicated, in favor of a myth, a story, a digression, leaps of thought, or a popular tale. But, reacts Latour, "these are allegories, which follow rules (*des allégories très réglées*). . . . Mine follow rules, too," answers Serres (*Conversations*, 73)!

For want of a novel vocabulary befitting the absolutely novel situation confronting Earth and humankind, Serres uses an allegory. The story of the interplay of strings within an inextricable knot expresses, however, a very *real* problem: How can the global object cohabit with the global subject? How can nature and culture live harmoniously under one roof? These are the questions of *The Natural Contract*. "[Humanity] discovers that it inhabits a global Earth that is the concern of our global science, global technology, and our global and local behaviors. This is the reason for the necessary synthesis" (*Conversations*, 152). It is a moral obligation; at stake is the postmodern survival of both humankind and nature.

Epilogue

In a postscript to his three-volume work, *Time and Narrative*, Paul Ricoeur presents "Conclusions" that may serve as a poignant example of the human dilemma vis-à-vis time. After what amounts to roughly one thousand pages of a profound, analytical reflection on time, fictional discourse, and historical narrative, the renowned philosopher carefully outlines the limits of language and representation when brought face-to-face with the aporetic of time. Words of caution, almost of reconsideration, creep into his conclusion which, he indicates in a footnote (in the French text), is the result of a rereading of his text a year after its completion. It seems as if time has had a prescriptive effect on the results of his research which now need to be redefined, conditioned, and circumscribed by a new meditation. Prescription, in the Serresean context of this term, is at work in Ricoeur's conclusion. Not only does he talk about the limits of his enterprise (Ricoeur, 241), but he also highlights the limits inherent in the problem whose solution he had set out to seek when he began the work (242–43).

Formulating a kind of epistemological principle of uncertainty, Ricoeur states that any meditation on time "suffers, quite simply, from not really being able to think time" (261). This failure is directly related to an intrinsic "*hubris* that impels our thinking [*notre pensée*] to posit itself as the master of meaning" [emphasis is Ricoeur's] (261). The failure to represent time poses an "embarrassing question" (243), a "troublesome question" (250), a "cumbersome question" (270) for the philosopher, three slightly different emphatic statements in the conclusion, which carry the same intensity of intellectual doubt.

Ricoeur's reservations concerning the possibility of a phenomenological outline of time, which he had hoped the interaction of fictional discourse and historical narrative could provide, echo the premise of *Genesis* to retrace the acquisition of knowledge to its pre-phenomenological Ur-noise. The multiple, the dynamical statue, and the instructed-middle are equivalent to "the mystery of time" to which Ricoeur refers repeatedly in his final pages. Similarly, the theme of *The Natural Contract*, which formulates a first and far-reaching application of the new way of understanding time, resonates in Ricoeur's postscript. The operating mode of prescription, in its most fundamental meaning, as we see it at work in the natural contract, is reflected in Ricoeur's description of the intimate relationship between time, subject, and object:

> Above all, only a philosophical story can situate the genesis of time beyond the distinction between psycho-logy and cosmo-logy, by forging the representation of a world soul that moves and thinks itself. Time is related to this hyper-psychological, hyper-cosmological "reflection." (263)

In the final paragraph of his "Conclusions," Ricoeur cautions against the *hubris* of reason, but adds that an outright rejection of reason invites "obscurantism" leading to deceptive cognitive processes engaged in an intellectual free-for-all. In a formulation that goes to the heart of his meditation on the "inscrutability" of time and the limits of discourse, Ricoeur confirms that

> [t]he mystery of time is not equivalent to a prohibition directed against language. Rather it gives rise to the exigence to think more and to speak differently. (274)

This statement reflects the issues at the center of Serres's cycle of writings on time. Ricoeur's appeal for a new discourse capable of broaching time in its mysterious working also bears a striking resemblance to Mitchell Feigenbaum's insistence that insights gained through recent dynamical systems theories "completely chang[e] the way to *know* something" (see p. 125). Both, philosopher and mathematical physicist, address the same issue, that is, that purely analytical methodology is no longer sufficient or even applicable. Possibilities for a new mode of thinking are forcing themselves upon the philosophical and scientific conscious.

Serres's writings for which this study has tried to provide a dynamical reading practice, recuperate Ricoeur's dilemma and enlarge its scope. The analytical approach that Ricoeur recognizes as woefully inadequate for an understanding of time was nonetheless his Ariadne's thread that guided him out of the labyrinth of reason's *hubris*. It was a necessary and liberating condition for a phenomenological outline of the "mystery of time." In contrast, Serres's overriding premise for the texts here discussed is a kind of bracketing of analytical methods. By leaving the straight path of philosophical and scientific inquiry, and opting instead for a "visiting" of circumstantial spaces, Serres reinstates a "method" which had been relegated long ago to the fictional, irrational, or emotive-intuitive spheres of perception. In none other than the Homeric text does Serres find the means to imply the essence of this neglected method and then to name it the "Ulyssean exodus." Only "implication," which we have come to recognize as a global strategy in Serresean thought, can see the method in the Homeric random wandering (*la randonnée*). Serres insists that explication alone is always stymied by innumerable investigations leading to theories but not to true understanding;

we end up no longer seeing the forest for all the trees. In a move that recalls his juxtaposition of Lucretian thought and modern scientific inquiry, Serres tells his reader the story of complex systems theory by telling the Homeric epic (*Les cinq sens*). It is a lesson in topology that is retold in *Statues*, and is applied to a double demonstration of the evolution of an inventive human being (*Le Tiers-Instruit*), and of a human social being in an utterly global sense (*The Natural Contract*).

Bruno Latour, Serres's conversation partner in *Conversations on Science, Culture, and Time*, presses him on the issue of the use of time in his writings. Latour is rather blunt when commenting on his habit of (re)aligning chronologically disparate historical events, followed by a concomitant breakdown of any linear sense of past and present. It appears to him as if Serres is "totally indifferent to temporal distance," and he adds: "One would say that for you there is no such thing as time. That everything is contemporary" (44). He pleads with Serres to shed light on his unconventional way of twisting past and present until they intertwine and thereby lose binary determinations of a before-and-after, then-and-now, or ancient-and-modern. Latour sees in the twisting of time one of the reasons for the intellectual jarring experienced by readers who judge Serres's writings to be obscure and difficult to follow: "This problem of time is the greatest source of incomprehension, in my opinion" (48). But Serres points out that in his writings the question is not one of playing modernity against a past that has been used and is done with. In other words, his juxtaposing of texts, authors, and myths into one time frame is not a juxtaposing of "times" expressed as past, present, and future, all lumped together. "I want to be able to understand time and, in particular, *a self-same time*" [emphasis is Serres's] (46). To illustrate the nature of this contemporaneity that seems to situate him outside of time, Serres points once again to Lucretius who,

> *in his own time*, really was already thinking in terms of flux, turbulence, and chaos, and, second, that *through this, he is part of our era*, which is rethinking similar problems. I must change time frames and no longer use the one that history uses. [emphasis is Serres's] (47)

The reference to *La naissance de la physique* combines several important issues which touch upon the essential elements our study hoped to demarcate and clarify. For us, it is significant that Serres uses the Lucretian example to enter into an extended debate with Latour on the problem of time (*Conversations*, 46–47). Lucretius' reflections on turbulence are analogous with present theories of turbulence in fluids and of chaotic complex systems in general. This permits Serres to find and place both in "the same [temporal] neighborhood" (57). But, and this goes to the heart of the "new time," time itself plays a crucial role within complex systems theories. As Prigogine and Stengers noted, we have to understand time as an operator; for most systems it is not a simple parameter expressed in measured increments that allows static systems to evolve linearly. Time is used here in a double sense, expressed by Serres's topological twisting of history, on the one hand, and by Prigogine and Stengers as an operator in the nonlinear evolution of systems, on the other. This double use of time allows Serres to recognize that two thought systems, separated by temporal distance and therefore judged as unequal, render equally valid concepts. Combined, these lead to a new notion of time, precisely that complex and "mysterious" time which allows Serres to place "temporally distant" systems in close temporal and logical proximity, in the first place. The argument is circular, but from a geometric, linear viewpoint only. From the perspective espoused in the texts we have assembled in this study, time is told (or fabled) as a topological complexity that Serres had already discovered in Lucretius' poem.

In *Conversations on Science, Culture, and Time*, Serres states this theoretically, pressed as he is by his conversation partner whose badgering underscores the importance of the problematic of time:

> Whether it's scientific hypothesis, on the one hand, . . . or, on the other hand, that of historicism, the two suppose that time develops in a linear fashion. . . . But time is in reality somewhat more complicated than that. You no doubt are familiar with chaos theory, which says that disorder occurring in nature can be explained, or reordered, by means of fractal attractors. . . . The development of history truly resembles what chaos theory describes. (57).

Serres continues by implying ahistorical time in terms of chaos theory concepts. What he calls "clarifying the theory of time" turns, however, into a series of remarks that demonstrate his mastery of a dynamical discourse practice: He recalls that the French language wisely refers to weather and time by the same word, *le temps*; he discusses percolation and filtration of fluids, a celebrated poem by Apollinaire, and a text by Péguy which presents Bergsonian views of historical consciousness (*Clio*); he points to numbers theory, the historical phenomenon of Nazism, different aspects of a folded and unfolded handkerchief, and shifting geographical distances; and he relates the story of a man whose age had been frozen in the same way that his body had been preserved in a glacier and whose aging sons mourn therefore a father younger than themselves (60–61). "It's precisely this biographical and philosophical bizarreness . . . that makes you so difficult to read," exclaims Latour in (mock?) desperation, playing the devil's advocate (61). Where Latour restricts himself to the role of moderator and partner in their discussions, William Paulson goes further and summarizes these oddities, similar to our summary reflections on Serres's "difficult" texts. Serres argues, he writes, that

> [t]he image of time's passage as *line* should give way to that of the *chaotic attractor*, a complex, enfolded, semirepetitive and yet non-self-identical

> trajectory, or to the topological figure of a wadded-up surface, in which points that are very distant in a two-dimensional metric can virtually coincide. [emphasis is Paulson's] (Paulson 1993, 36n4)

The image of time as a chaotic attractor is mirrored through different models in the texts discussed in the previous chapters. It allowed us to connect the many apologues to a complex picture of variabilities out of which we slowly distilled a discursive chaotic attractor for which we read a particular trajectory in *Statues*. In this text, we encountered Serres's portrait of time as a dynamical experience. However, the impression of a linear progression through Serres's texts is inevitable, given that a study such as the present one cannot present its object of reflection in a chaotic pattern, or itself would entail the necessity of a rereading. Any linearity is, however, purely heuristic and should not displace the recognition that the texts here assembled are discursive portraits of a complex dimension that marks human life. We must also emphatically repeat that the selection of these texts is not an arbitrary choice; neither does it indicate an essential division within Serres's epistemological writings as a whole. To the extent that the philosopher speaks of a cycle of books when he refers to the writings which deal overwhelmingly with theories of communication, we have referred to a cycle of texts on time. If, therefore, stages appear to exist in the evolution of his writings, they should be understood to imply unique trajectories expressing the evolution of various aspects of human reality, but together marking or tracing out the same region of great complexity, which encompasses the subjective intellect and objective reality in an epistemic setting that is global in a nonlinear fashion.

Josué V. Harari and David F. Bell characterize the *Hermès* series as an investigation into the messenger god's "displacements . . . related to the problem of space" (*Hermes*, xxxiii). Serres himself recuperates this problematic into the larger setting of a cycle of writings on communication. We, on the other hand, have discerned a grouping of his texts on time. Yet the notions or models of space expressed through communication (passages), and those of time presented in numerous apologues, aphoristic stories, and textual complexity, are not essential divisions within Serres's epistemological work. Nor are they linearly opposed facets concretized in the various discursive units we recognize in each of Serres's books. Rather, they are truly understood if and only if we see their equivalence on a deeper level. Serres argues that the biased attachment we harbor for *linear* time, with its corollary of progress, is no less "simplistic, banal, and naive" than ancient beliefs which placed the Earth at the center of the visible world, our galaxy in the middle of the universe, and ourselves as the measure of all things:

> The curve traced by the idea of progress thus seems to me to sketch or project into time the vanity and fatuousness expressed spatially by that [pre-

> Keplerian] central position. Instead of inhabiting the heart or the middle of the world, we are sojourning [today] at the summit, the height, the best of truth. (*Conversations*, 48–49)

Serres attributes this false sense of permanence to the privileged position modernity gives to the present.

Both models, spatial centrality and temporal progress, are erroneous in Serres's view. There is a deep analogy bordering on equivalence and identity in his arguments which address the problem of space and those which seek a new understanding of time. It is not an arbitrary discursive event when, early on in *Le Tiers-Instruit*, Serres reflects on the Keplerian revolution with its decentralizing twisting of our world into an elliptical construct of space. The notion of "elliptical space," in which the human center is literally upstaged and pushed aside, had already been expressed in the Serresean discourse as a world of passages validated by Hermes, and as an "elliptical" epistemology modeled in the figure of the parasitic intrusive third (*The Parasite*). In *Le Tiers-Instruit*, the ellipsis of space is described as the twin of an "ellipsis" of time, where epistemological linearity gives way to an offbeat movement whose chaotic trajectory results in the portrait of the inventive *I*. The idea of time as the carrier of progress, with its attendant dark side of monumental and often deadly failure, is replaced by the notion of unpredictability with its promise of inventive force.

However, there is another side to the discourse on time and its deep-structured ties to Serres's work as a whole. This is an aspect that is perhaps most profoundly connected, *in the paradigm of the statue*, to the overall vision of the epistemologist and "epistemo-critic." (By the same token, it gave the initial impetus that produced the present study.) But before this aspect can be formulated, we need to summarize once more the foundations out of which it evolved. Serres's books on communication model the problem of space through passages where the messenger god of communication becomes the paradigm for a field model of space as fluidity or fluctuation. "New" knowledge lies in the inbetween, or the passage between fixed points of knowledge. The wily god offers knowledge in the never stable region of passages. The parasite is the more human incarnation of Hermes; his intrusive and subversive work is circumscribed in the function of a "quasi-object" which is nowhere at home, yet is everywhere present. But irrespective of the many different angles from which Serres approaches the problematic of space, his demonstration always includes the "anti-model" of the statue as the paradigm of *stans*, space as a static or thetic construct. The statue of Molière's Commander in the last chapter of *Hermès. La communication* is the first remarkable figure with which Serres models the full authority of the stable and immutable Law, and against which he pits Dom Juan's shifty logic and manoeuvering. The Commander's shadow looms large in Serres's texts on communication where any argument, pointing to stagnation, rigidity, stability, repetition, thesis, unitary, or binary constructs, is ultimately expressed in analogies involving statues or statue-like phenomena. The statue as the incarnation of absolute immobility, or as rigid perfection no longer in need of inventive improvements, haunts the writings of the philosopher of communication. It bursts forth now and then, in the texts we have gathered, as we have indicated at several instances in this study. Ultimately, the statue figures as the epistemologically created obstacle that blocks the passages, just as the great Alexander screened the sun in front of Diogenes. It is the specter of death, the totality of stability and of absence of variability. It is Serres's model of evil.

It is therefore the more remarkable that the statue becomes the vehicle for Serres's discourse on a "new" ahistorical time. The statue of Aphrodite Anadyomene of *Genesis* leads to the hopeful final words of renewal in *Les cinq sens*. These grow into a jubilant crescendo in *Statues* where the Ulyssean meandering and the knowledge of circumstances become a dissipative understanding traced out by the Statue. At the heart of the most static and stagnant of all his discursive tropes, Serres discovers a complexity and an inventive power so great that he can speak of "collective immortality" by which he means a time that is never subject to historical references and their inevitable destructive consequences. From our perspective it is significant that he presents this "chaotic" statue in a discursive double header: The first part of *Statues* is subdivided into two passages, "The First Foundation" and "The Second Foundation." The First Foundation is a direct reference to *Rome: The Book of Foundations* and gives a discursive portrait of the statue as *stans*. It is the static statue's last stand, so to speak. Its last hurrah, however, is also a shocker unless we accept Serres's topological time: Super-modern space missions with all their technological refinements repeat ancient rituals of human sacrifices to the gods; only the accidentals or the trappings have changed. The static or thetic is the giant monument (*statue*) to humankind's worship of death. Death is as much the *telos* of our high-tech era as it was of ancient mythology.

The Second Foundation is a detailed account of Serres's inexorable descent into the epistemological statue where he discovers a chaotic complexity so rich that it explodes all tenets of the first foundation. The static monument changes into a topological field on which time plays out its multiple inchoative moments, and where the absence of repetitive instants invites continuous new findings with the power of invention. Serres, in fact, acts out in his own discourse the story of the pious peasant of myth who made many offerings and sacrifices to his idol god. Enraged that the latter does not answer in kind, he smashes the wooden statuette and discovers a fortune in its hollow that far outweighs what he had prayed for in vain (*L'Hermaphrodite*, 114). With the help of Jules Verne's story of the astronauts and the dog Satellite, Serres succeeds in breaking open the statue that had hounded him as much as Molière's Commander had pursued Dom Juan, and against which he riles with the same spirited and parasitic interference as the great lover of women and words. The treasure he finds is death as a matrix and generatrix, as a companion-partner to the subjective intellect and the objective world.

We have elaborated on this discovery in chapter 4 dedicated to *Statues*, where we have read of vast epistemological implications for a kind of "novel, secret, extraordinary" stability already alluded to in *Genesis* (see p. 42). For the first time, the stance is not pitted against flux or multiple variability, but is instead allied with time in a chaotic evolutionary process. In *Les cinq sens*, Serres expresses this paradoxical pairing of the stable and the unstable, of order and chaos, with an invitation to have both sit down together at the banquet of love. In *Statues*, the pair is enriched by a third, with the addition of death; the three then become one in an undecidable process. We have traced this process with concepts of nonlinear dynamical systems theory by which scientists today present "portraits" of order and chaos bound together with reference to time as an operator. *Statues* brings together order and chaos in fuzzy bounderies where the dynamics of subject and object are no longer distinct, and where time plays its operating role in the notion of generative death. In this way, a portrait emerges of "unstable stability," which we have referred to as a "dynamical statue" simulating a "new" knowledge that unites subject, object, and generative death.

Serres discovers that the statue as a figure of stability has always covered up death as absolute disorder, in the mistaken belief that "unnameable death " (*Bossuet*) could thereby be entombed, forgotten, and, for good measure, named into something stable. With this uncovering, he obtains a true picture of all human cultural endeavor. With it he gains the knowledge of a universal epistemological cover-up or lie that inevitably leads to dangerous consequences. This is the problem of evil (*le problème du mal*) against which the Serresean discourse formulates models for a new morality. *Statues* and *Le Tiers-Instruit* present the means to reveal the fetish or idol-statue for what it is, and to look death squarely in the face, so that we, like the ancestor standing in awe before the ancient sequoia tree (*Detachment*), may once again understand death's generative powers and regain our will to resurrect into a "collective immortality."

Statues brings into focus what Serres has been laboring to imply in his "many books" since *Hermès*, to find an antidote to the static which is the core of evil. It sharpens the image of the dynamical statue which always lurks in the wings whenever Serres discusses the statue as *stasis*. I regard *Statues*, therefore, as a key text in the series of texts on time, where generative death is analogous to topological time. But this text is not the only one to speak of death. The year that *Statues* appeared in print saw also the publication of *L'Hermaphrodite. Sarrasine sculpteur*. Ostensibly a literary commentary on Honoré de Balzac's novella "Sarrasine," this monograph presents the triad of subject, object, and death within a fictional setting that itself models its tale of determinism and undecidability around a statue.[1] The sculptor's enraged blow that narrowly misses the statue is the only honest gesture in the tale, because, if successful, it would have revealed a truth: the chaotic foundation of human action. Out of this disorder, mirroring the sexual ambiguity at the core of both Balzac's story and his narrator's story within the story, Serres sees the novelist spin a tale about death begetting a statue, begetting value, and begetting a family's good fortunes. The episode of the dog Satellite taken from Jules Verne's story and retold in *Statues* to demonstrate the Second Foundation is the companion piece to *L'Hermaphrodite*. With Verne's help, *Statues* converts into a global setting what in *L'Hermaphrodite* is essentially the story of a particular family history, a local founding that is typical of many of Balzac's philosophical stories. It marks the Balzacian statue as a singularity within the vast cultural background that is ours and that we have modeled, and still model, in statues and statue-like phenomena. Serres reads "Sarrasine" as a story that models neither an exclusion (as the castrate Zambinella may indicate in a conventional reading) nor plain disorder (through the sculptor's murderous rage or the deathly specter of the old gentleman who haunts the elegant and festive ballroom of the family Lanty). Rather, it is the story of "over-abundant androgyny: we have to conceive of an extravagant, extraordinary, overflowing, hyperbolic Hermes (*il faut concevoir Hermès comblé*)" (*L'Hermaphrodite*, 88). This wonderful portrait of Hermes-in-excess or emergent Hermes is contrasted with the stagnant repetitions of literary analysis in general that picks apart a given text in the hope of unlocking a plurality of meanings, leaving nothing but "erroneous calculations."

We, however, should see in *Hermès comblé* the winged Hermes we briefly encounter in the final pages of *Les cinq sens*, the Hermes-statue breaking out of its own static order and embracing the chaotic fullness of Hermaphrodite. Less poetically than does Serres and more prosaically, we call it a "dynamical statue." The androgynous figure of the parasitic third, called a "hermaphrodite" in the last paragraph of *Statues*, is actually vibrantly dynamical, all-inclusive, overabundant, overflowing, excessive—without ever being absolute or static. *Hermès comblé* is the lesson of *L'Hermaphrodite*. With it Serres can fabricate ("poiein") shortly thereafter a natural contract and an instructed-middle. As if it were at once a preface and a postscript to the dynamical new statue we meet in *Statues*, *L'Hermaphrodite* fabricates out of Balzac's "Sarrasine" the portrait of a new knowledge being born of generative death:

> I see, I know, I experience death as a beginning, and not as an end. Far from destroying us, death produces us. Behind us, in us, death pushes us into culture, customs, words, thoughts, human-ness, and humane-ness. . . . Time no longer flies toward death but flows from it. Let us leave the dead to bury the dead. Now I am awaiting something akin to the Resurrection for our modern collective death which appeared barely a half century ago and which, behind us and in us, disturbs life and history today. (113)

It is the search for a new concept of time that has brought Serres back to the statue, only to loosen its static bonds and thereby to destroy its absolute, lethal hold on our life as we live it today. The dynamical chaotic statue no longer hides death but offers its generative powers for a renewed knowledge. That science concentrates more and more on complex, disorderly systems and adjusts its methodology to account for a new notion of time as an operator, only reinforces the urgency of Serres's plea to his reader to account in our epistemological endeavors for a new notion of death as generatrix. Both time and death are two aspects of the same experience, the union of subject and object in the liberating experience of "collective immortality."

The present study of Serres's texts on time has gathered writings that prepare and advance the Second Foundation, which receives its *explicit* full form for the first time in *Statues. Le second livre des fondations*. The first book of foundations addressed the origins of historical time using the example of Rome; the second reveals ahistorical time and with it the creative force of death as the only valid response to the problem of evil. "Collective immortality" is the promise or expectation issuing from this philosophy of time. *Les Origines de la géométrie* (1993) is Serres's Third Book of Foundations in which he recalls the first two and speaks of a "preliminary philosophy of time" (15). But the third book hopes to do away with all tombs, that is, with all statues. It presents yet "another time founded on the absence, the opening or the forgetting of any tomb. Invention works against death: and perhaps toward immortality?" (60). Serres the geometrician reflects on the origins of geometry to arrive once again at the fluid foundations of knowledge, linked to the notion of time as collective immortality. It should not come as a surprise that *Les Origines de la géométrie* often mentions the texts we have discussed and cites many passages from *Detachment*. Collective immortality is one of the crucial notions in the latter and is in essence Serres's metaphor for a new morality that finds its field of action, again metaphorically, in the topological space of angels, dragon kites, and lovers. The publication of the book on the origins of geometry coincides with that of *Angels: A Modern Myth* [*La Légende des anges* (1993)]. The latent connection between these two works, so different one from the other in discursive style and material composition, can easily be recognized if the reader recalls the writings on time and sees with Serres an *implied identity* between *the topological space* that we can only reach in an angelic state of detachment, and *inventive time* as a promise of collective immortality. *Le Tiers-Instruit* and *The Natural Contract* have shown us how this promise can be modeled on the individual and the cultural-natural levels.

In *Conversations on Science, Culture, and Time*, Latour is almost incredulous when Serres explains that Hermes has changed into a multitude of angels (117–119). "This is not going to clarify things for the public. You are supposed to be offering us clarifications here!" (118); and "this is a theological quarrel!" (119), exclaims Latour impatiently. But Serres insists that Hermes is still too much the figure of a god whom we moderns may mistakenly turn into a stable system. Therefore, Hermes the message-bearer changes into Hermaphrodite, the undecidable one, he then becomes the fuzzy region of multitudes of angelic messages. Today he is implied in "their cloud, their whirlwinds. I was about to say their chaos, since their collectivity is similar to it. . . . Pure multiplicity" (119). From Hermes to the angels, Serres travels passages that seem to lead him astray. But the present study tried to demonstrate that this odyssey is most natural for our times. From its perspective, *Conversations on Science, Culture, and Time* enlightens but does not explain. The texts it has been reading are the key for an enlightenment where explanations become superfluous when implications model reality.

To the question why one should today be interested in angels, Serres replies that "our universe is organized around message-bearing systems, and because, as message-bearers, they are more numerous, complex and sophisticated than Hermes, who was only one person, and a cheat and a thief to boot" (*Angels: A Modern Myth,* 293). Any topology for a renewed morality needs the variability and *mélange* of angels. "Perhaps what I was writing all along was an angelology" (*Conversations*, 118). Whether we read *la belle noiseuse*, the dancer in the sky, the seraphic joy of the human full of sensate experience, the strange attractor of generative death, the strange attraction of the inventive *I*, or the contractual union of subject and object on the global level, this study has tried to read such an angelology into their chaotic multiplicity of messages and passages.

Notes

Introduction

1. Classical scientific methods cull models from idealized environments and eliminate unquantifiable elements that may disturb the universal value of a given model. It is therefore a structured *representation* of naturally occuring phenomena, but is not "natural" in itself.

2. The individual titles of the *Hermès* series, and even more so *The Parasite*, reflect Serres's concentration on information theory.

3. The notion of *passing* acquires a defining meaning for the pedagogical argument Serres puts forth in Le *Tiers-Instruit*. This will be discussed in detail in chapter 5.

4. In chapter 1, the term "posit" will rapidly tend toward the more characteristic "imply" at work in Serresean discourse. The passage from posit to imply describes the shift from the parasitic third to the multiple, to complexity.

5. See the interview granted to *Le Nouvel Observateur*, no. 1513, (November 4–10, 1993): 54.

6. I derive this formulation from Stephen Kellert's "dynamic understanding" (*Complexity*, 114), which he explains as "research into nonlinear systems [that] provide a new kind of understanding, which could lead to a new conception of nature that accepts randomness and contingency" (*Complexity*, 116).

1
Time Promised: Reading *Genèse*

1. In old French, "noise" (*nwaz*) meant noise, uproar, quarrel, din, furor. It is rarely used in modern French other than in the expression *chercher noise*, to pick a fight. Throughout this study, the French "noise" will be identified with quotation marks. No marks will be used when the English noise is indicated.

2. A measure of Serres's more disorderly approach is to be found in the decreasing application of models like "the multiple" and other fuzzy metaphors to come, by critics and theoreticians of literature and other areas in the humanities.

3. The relationship I am trying to establish between Serres's epistemological findings and concepts of complex dynamical systems theory is isomorphic in the scientific-

mathematical sense of this term, as Douglas R. Hofstadter describes it: "The word 'isomorphism' applies when two complex structures can be mapped onto each other, in such a way that to each part of one structure there is a corresponding part in the other structure, where 'corresponding' means that the two parts play similar roles in their respective structures." *Gödel, Escher, Bach* (New York: Vintage Books Edition 1980), p. 49.

4. A poignant reference to this Ur-silence which is Ur-noise is made in *Statues*, subject of chapter 4 of this study.

5. Stéphane Mallarmé, "Le Mystère dans les lettres," *Oeuvres complètes* (Paris: Gallimard, 1945), pp. 375–87; in particular pp. 386–87.

6. The Mallarmean passage, which figures so prominently in the first pages of Derrida's "La double séance," is itself headed by a sort of double title given by the poet: "Lire—/ Cette pratique—." Only then follows the passage containing the remark on the title. What Derrida left unsaid, since his aim here is writing (*écriture*) as one of the key terms of his philosophical work, becomes significant for our reading of *Genesis* at this juncture: The question of the title as "Ur-noise" relates to the practice of reading (*lire*). One of the objectives of our study of Serres's writings is a dynamical reading. The practice of reading the back cover of *Genèse* in the fashion of a Mallarmean reading practice may well be the first step toward a dynamical reading.

7. In a recent interview and as a response to an interviewer's questions about his unorthodox disregard for accepted scholarly practices, Serres explains that quotations, when inserted in a text, are there to speak for themselves and do not need to be identified. Implicit in this strategy is the assumption that his reader will recognize the source of the borrowed passage. In the texts discussed in the present study, footnotes are extremely rare and then do not always serve to indicate other textual sources.

8. The dynamical union of culture and nature will much later become the subject of *The Natural Contract* discussed in chapter 6 of this study.

9. W. Paulson disagrees with the accepted analysis of the appearance/disappearance of the sobriquet *la belle noiseuse* in and from the various editions of Balzac's text ("Pour une analyse dynamique de la variation textuelle: *Le Chef-d'oeuvre* trop *connu*." *Nineteenth-Century French Studies* 19, no. 3, (Spring 1991): 404–416). However, Serres's reading addresses the conventional interpretation of Balzac's intentions and is therefore valid for our reading within the wider Serresean context.

10. See "Michel Serres: In Search of a Tropography," *Chaos and Order*, N. Katherine Hayles, ed. (Chicago: The University of Chicago Press, 1991), pp. 278–98.

11. Increasingly, theoreticians of complexity focus their attention on these thresholds of phase transition occurring in dynamical systems, as potential regions of richly creative activity.

12. I prefer the adjectival form to "possibility," because it translates the Serresean epistemological use of adjectives as well as the purely literal form of "le possible" used in the French original.

13. The 1958 edition of *Little and Ives Webster Dictionary* gives a formulation that is even more interesting in the context of our reading: "Land permanently exposed by retreat of the sea."

14. Serres's toying with *mur* and *murmure* is not quite as ingenious as it appears at first and has a precedent. In 1785, the easily aroused Parisian crowds found something to vent their anger on when new city walls were built to facilitate the business of tax

collectors. Someone elegantly lampooned this indignation with an Alexandrine verse that remains well-known to this day: "Le mur murant Paris rend Paris murmurant" [The wall that surrounds Paris makes all of Paris grumble] (Curtius, 205).

15. "Thesis" and "stasis" are interchangeable terms in Serres's discourse which is steeped in references to Latin and Greek etymological roots. Both refer therefore to a stationary state or repose and frequently have negative connotations of stagnation, immobility, or rigidity in Serres's discourse. The adjectival forms "thetic" and "static" share the same meanings. I want to emphasize in particular that here and throughout this study I use "static" in the sense of non-dynamic or stationary. It should not be read substantively as noise or static which is an interference phenomenon in communication systems. As an important concept in information theory, static (noise or white noise) plays, of course, a major role in many of Serres's earlier texts.

16. The biblical quotes are taken from *The Jerusalem Bible* (Garden City, NY: Doubleday and Co., Inc., 1966).

2
Time Immortal: Reading *Détachement*

1. Boustrophedon writing, literally the writing of the ox tracks, is an early primitive Greek writing characterized by lines that go from left to right and right to left alternately, thus forming a continuously undulating line of script. Serres uses this writing as a model for his reading of early Roman history (*Rome* 19–45). As he once explained in a lecture, boustrophedon writing enables him to read the proverbial lines between the lines.

2. In a synergic gesture so often encountered in Serres's writings, the discursive subject of the peasant and the authoritative voice of the discourse collapse into one. This produces an epistemological equivalence between the activities of the peasant and those of the writer.

3. Petrus Christus whose mentor is thought to have been Jan Van Eyck was steeped in this religious tradition. He and his wife belonged to the Brotherhood of the Withered Thornbush (*la Confrérie de l'arbre sec*), a lay organization whose spiritual aims were similar to those of the important Brotherhood of the Rosary.

4. There are three sets of five mysteries each. The rosary commonly referred to is actually any one of the three sets and associates ten Hail Mary's with each mystery, creating a rosary circle of fifty Hail Mary's. The three sets together form a string of 150 Hail Mary's equivalent to ten times the Marian number of fifteen.

5. The influence of Rousseau on Serres's work is discussed in more detail in chapter 6.

6. See Ilya Prigogine and Isabelle Stengers, *Order out of Chaos* (Boulder, Colorado: Shambhala Publications Inc., 1984), 269, for a more detailed explanation of this notion used in discussions on topology and complex systems theory.

7. Baudelaire's poem ends with an exhortation for the ultimate revolt: Not only is Cain's murderous crime neutralized by the passage of time, in a process of prescription, but Cain's race is encouraged to tend to its unfinished business and force God himself to his knees. Such dramatic and sacrilegious statements occur frequently in Baudelairian

poetry. I am not suggesting that Serres's critique of our cultural icons implies such an extreme gesture. But "Abel and Cain" presents a clear picture of the workings of time as an ahistorical process which allows for the possibility of destroying fetishes and, by implication, eliminating other areas of competitive rivalries. From this perspective it functions like the epistemological critique presented in *Detachment.*

3
Time Empirical: Reading *Les cinq sens*

1. In German, "to praise" (*preisen*) and to "part with" or "give up" (*preisgeben*) are based on the same root "*preis*" and therefore render Rilke's comment the more allusive and, above all, more elusive. It exposes the inherent impotence of language (praise) to reveal the objective world in its full immedicacy (give it over to us by parting with it).

2. The proofreader stands metonymically for all relationships of competition and judgment, and represents the particular competitive encounters in the world of the writer and academic.

3. Serres is convinced that his writing will be fabulous or miraculous enough to allow the subject-reader to break through language and reach immediate reality. Therefore he protests the poetic label often attached to much of his recent work, because it will not do justice to his writing, in the sense of Rilke's comment about the limits of poetic language. Serres insists that his writing is a "fabrication," in the meaning of the Greek verb "to fabricate" as an act of "making" (see introduction). In this sense, and if it is truly and radically poetic, his language creates conditions of possibility for immediate reality to be experienced by the reader.

4. The most difficult and complex text among all of Serres's writings, however, is no doubt his study on aesthetics, *Esthétiques: Sur Carpaccio* (Paris: Hermann, 1975).

5. In *A Cognitive Theory of Metaphor* (Cambridge, MA: MIT Press, 1985), Earl R. MacCormac explores possible semantic parallels between the ambiguous structure of metaphors and fundamental terms of fuzzy set theory, pioneered by computer scientist and mathematician L. A. Zadeh.

6. The authors give a simple example of such a linguistic variable: "If age is treated as a linguistic variable, its values might be young, old, middle aged, very young, not very old, and so on, with each value interpreted as a label of a fuzzy set. This concept of a linguistic variable serves as a basis for computing with words. This is at the heart of most practical applications of fuzzy logic . . . [when] a mathematical model would [be] too complex to be manageable" (Coveney and Highfield 1995, 75).

7. The word *vair* is derived from the latin *varius* meaning variegated and aptly characterizes the varied colors as well as layers of short and long hair that distinguish squirrel fur.

8. The best-known visual of fractals is no doubt the Mandelbrot set which is a generalized form of what are known as "Julia sets." The intricate beauty of these computer-generated graphs is mathematically known as fractal geometry which provides a practical tool for research in many fields of the natural sciences. See Peitgen, Gleick, and Mandelbrot for further explanations of this fascinating topic.

9. By coining the expression *la geôle d'ivoire*, Serres combines the image of a prison (*geôle*), often associated historically with a dungeon or tower, with an ivory tower (*ivoire*), image of the academic world dominated by the myriad forms of representational activities. This is a good example of Serres's poetics at work: while fabricating an image by slightly twisting an accepted and well-known trope, he summarizes in one unique expression the analogy of the poisonous drug and the word, of the physical prison and the house of language. Upon this double analogy hinges the demonstration of the significance of the story of Socrates' last hours.

10. Serres implies a link between Socrates' love of words and his clumsy and ugly physical appearance that tradition and satirists of antiquity attribute to him. Correspondingly, true beauty is associated with the immediacy of sensate experience.

11. Serres describes the same process of "em-boxing" in theoretical fashion in "The Origin of Language," where he speaks of the transformation from noise to information by way of "correctors-in-series." The single black box disappears under the form of a series of Russian dolls. The series becomes the information-processing organism (*Hermès IV*, 268).

12. The Latin link between sapid and sapient is tenuous, although "sapor" can mean the sense of taste as well as taste of style or distinction (see *Oxford Latin Dictionary* (New York: Oxford University Press, 1982).

13. Serres never hyphenates the two sides of circumstance, and for good reason, as our discussion in chapter 4 will show. I have introduced the hyphen here out of a purely heuristic consideration and shall abandon it once the dynamics of the word are established. However, in *Conversations on Science, Culture, and Time*, Serres hints at the importance he attaches to the hyphen in the "tiers-instruit," from which I borrow to justify my momentary use of circum-stance.

14. See Prigogine and Stengers, *Order out of Chaos*, for a more explicitly scientific language of integrable systems, and for the impact the *recognition* of non-integrable systems had and still has on scientific and mathematical research today.

15. In addition to the gospel attributed to him, the apostle John wrote letters to the faithful which are now part of the New Testament. Their emphasis on love, as well as legends surrounding the youngest of the twelve apostles, have given him the epitaph of "apostle of love." In his writings on religion, Gotthold E. Lessing, philosopher, writer and critic of the German Enlightenment, comments on this aspect of the evangelist's message and leaves his reader to wonder if Christianity might have acquired a different character, had it not privileged Paul's stern teachings to the extent that it did. (See in particular "Das Testament Johannis," *Gesammelte Werke*, ed. Paul Rilla Vol. 8 (Berlin: Aufbau-Verlag, 1968).

On the other hand, Serres comments in great detail on Dom Juan the lover, in *Hermès.La Communication*, where he underscores the parasitic nature of the lover who pits his chaotic work against the statue of the Commander, emblem of stability and order.

In *Les cinq sens*, Serres ingeniously equates John the apostle of love and Dom Juan the lover, with the intention of describing the nonlinear and circum-stantial feast of the senses as opposed to the feast of the statues and *stasis*, allegorized in Peter the Rock and the specter of the Commander.

16. My translation of this quote is, at best, loose. In the original, the text reads: "Je rêve d'écrire un troisième banquet où la statue, vengeresse, accepterait de boire du vin fin avec qui a suborné sa belle."

17. In the last chapter of *Les cinq sens*, Serres devotes a lengthy passage to the "healing" properties he (still) finds in his own culture by which he means not only French but also Gascony traditions. These, he characterizes as a "culture of feminity" (365).

18. One of Serres's most recent works (as of this writing) is aptly entitled *La Légende des anges* (Paris: Flammarion, 1993). He underscores the role of Hermes encapsuled in modern techno-mythological tableaus which vividly reflect the role of the angel-poet who announces the coming of a new wisdom.

19. The substantive of *gésir*, to rest or lie, is *gisement* referring, among other meanings, to the nautical sense of bearing. In the context of the *Odyssey*, the bearing of islands takes on a particularly rich and apt connotation.

20. Serres was the keynote speaker at a colloquium, "Epistémocritique et cognition," held in Paris in March 1992. Its program centered around considerations of postmodern challenges to accepted epistemological practices. The proceedings of this colloquium were published in France. *TLE*, nos. 10, and 11, (1993). In the United States they were published in *SubStance* 71/72, vol. 22, no. 2, and 3 (1993).

21. Serres establishes a similar analogy in *La naissance de la physique*, suggesting a scientific reading of Lucretius' work that would complement and reorient its traditional poetic interpretation.

4
Time Dynamical: Reading *Statues*

1. This chapter is an expanded and extensively modified version of an article published in *SubStance* 71/72, vol. 22, nos. 2, and 3 (1993).

2. In spite of his enthusiastic support for the new "science of Complexity," Roger Lewin, for example, balances his account of the historical inception of this branch of nonlinear dynamical systems theory with many cautionary and/or skeptical commentaries by scientists in various fields of research. In this respect, his *Complexity: Life at the Edge of Chaos* (New York: McMillan, 1992) is clearly a more objective presentation than James Gleick's unabashedly triumphant *Chaos* which, however, has the distinction of having brought the "new science" into the limelight and popular conscious.

3. For an in-depth introduction to dynamical systems theory, the article "Chaos" by James P. Crutchfield, J. Doyne Farmer, Norman H. Packard, and Robert S. Shaw still remains one of the best and most readable sources of information for the nonscientific reader. Also very useful for more global applications of chaos theory are James Gleick, *Chaos: Making a New Science* (New York: Penguin Books, 1987); and N. Katherine Hayles, *Chaos Bound: Orderly Disorder* in Contemporary Literature and Science (Ithaca, N.Y.: Cornell University Press, 1990).

4. Graphs of such attractors can be found in James P. Crutchfield et al., "Chaos," *Scientific American* 255.6 (December 1986): (49–50), and James Gleick, *Chaos: Making a New Science* (New York: Penguin Books, 1987), (136–43).

5. This and all following quotes from *Statues* are my translations. The original French text reads: *l'immortalité collective* (22).

6. ". . . les techniques efficaces et les technologies culturelles. Les premières construisent le monde, les secondes construisent le temps" (23).

7. ". . .[*la mort*] plane . . . au-dessus de nous, au sens global de l'humanité entière" (23).

8. The predictions for the Shoemaker-Levy 9 comet's course and collision with Jupiter in July 1994 give testimony to the constancy of Newton' laws and their enduring practical applications, a fact we must not forget in the age of the new science of chaos and complexity.

9. In *The Natural Contract*, we read such a point attractor on a global scale in Serres's portrait of ecological destruction (see chapter 6).

10. In the French original, Serres borrows the image of the bumblebee from a line by Verlaine, to express haphazard and disorderly movement: *la guêpe ivre de son vol fou.*

11. A reproduction of the Lorenz attractor can be found in Gleick, *Chaos*, 50. A very comprehensive visual collection of strange attractor graphics is presented in H.-O. Peitgen and P. H. Richter, *The Beauty of Fractals: Images of Complex Dynamical Systems* (New York: Springer-Verlag, 1986).

12. This expression occurs in the following context: "Whether we are considering things, systems, ideas, or human beings, closure amounts to sterility, but any fertile action must pass through openness. This reversal of values parallels the transformation of space." [. . . , qu'il s'agisse des choses, systèmes, pensées ou personnes, la clôture équivaut à la stérilité, toute fécondité passant par une éclosion. Ce renversement des valeurs suit la transformation de l'espace"] (p. 59).

13. Serres had already dedicated a book length study to Verne's texts, *Jouvences: Sur Jules Verne* (Paris: Minuit, 1974). In it, he used a scientific approach to demonstrate a common *savoir* shared alike by literary and scientific-mathematical discourses.

14. "Entre Barbicane et ses camarades, le chien, en excès, doit passer par la trappe pour que le groupe puisse survivre: nous avons assisté au rite de constitution du sujet collectif" (p. 42).

15. "Aucune instance dans cet espace n'a donc encore pris son rôle: le chien mort est objet ou sujet; l'observateur dans sa bombe de mort reste enveloppé dans un objet, comme je le suis dans le livre et les lignes.

La mort stabilise-t-elle les rapports du sujet à l'objet? L'objet, à son tour, stabilise-t-il les rapports du sujet à la mort? Individuel ou collectif, le sujet stabilise-t-il les rapports de la mort et de l'objet, au point qu'il ne sache pas décider?

Cette triple stabilisation définit le statut des statues, dures comme des objets, corps de sujets morts, conservés ou revenus" (pp. 42–43).

16. Instead of "generator," I prefer to use the expression "generatrix" (without necessarily signaling its mathematical connotation in English). Since death is a feminine noun in French, Serres refers to "la mort génératrice" and understands it as a matrix, a life-giving element. "Generatrix" comes closer to conveying this meaning.

17. "Les objets se retirent au fond du monde ainsi que les sujets au creux de leur boîte, à distance astronomique les uns des autres, laissant le cadavre à petite portée constante des murailles de l'habitacle" (p. 45).

18. "La deuxième fondation, dont nous ne savons pas si elle précède ou suit la première, en découle ou l'approfondit, met en relation la mort avec l'objet. . . . l'autre [*fait voir*] la face illisible et silencieuse, invisible, d'une instance fondatrice, qui n'a de

nom dans aucune langue et qui assemble les instances que nous découpons sous les trois noms d'objets, de mort et de sujet. Ce gisement fondamental unit ce qui gît dessous, ce qui ci-gît et ce qui gît devant. Objectivant le sujet, la mort lui donne l'objet à condition qu'il le travaille. Comment nommer ce gisement, cette instance stable, sinon une statue?" (p. 48).

19. The unnameable designation of death that appears in Serres's text echoes a famous passage in one of Bossuet's *Oraisons Funèbres*, the eulogy for Queen Henrietta of England.

20. "La mort fait le rapport stable des sujets et des objets" (p. 45).

21. "Or les interprétations du mythe, y compris la mienne propre et le calcul savant, ne parlent que de la scène et du héros, coupable, malheureux, devenu forçat. Nous ne voyons jamais que nous-mêmes. . . . Et si pour une fois nous regardions la pierre, invariablement présente sous les yeux, objet entêté gisant devant?" (p. 302).

22. "Cette pierre roulée, nous avions perdu la mémoire qu'elle était à la fois notre antériorité immémoriale et notre totale postérité, savoir le genre humain. Les pierres forment les squelettes des morts et les semences de l'avenir" (p. 304).

23. "Ce travail sans cesse repris, où paraît cet invariant, permet, enfin, ici, de tracer la trajectoire de la pierre que nul, jamais, ne dessina" (p. 310).

24. ". . . le chaos peut se définir comme une variété où le désordre empêche l'apparition d'un lieu. Il ne fournit pas de repère ou nous ne pouvons pas y en discerner un. . . . A la question où?, le chaos ne répond pas" (p. 333).

5
Time Inventive: Reading *Le Tiers-Instruit*

1. This chapter evolved out of my research for an article published in *Synthesis* 1, no. 1 (1995): 73–86. (The University of Tennessee Press). It constitutes an expanded version of the argument pursued therein.

2. Throughout this chapter and following Serres's instructive lead, the "tiers-instruit" is understood as a *qualitative* alterity to the "tiers exclu." Since the latter is in English the "excluded middle," I shall consistently operate the same analogy and use "middle" for "third" (tiers [masc.], tierce [fem.]), unless it becomes necessary to do otherwise if a given argument is built on an explicit image of the third. Therefore, the "tiers-instruit" is the instructed-middle, the "tierce place" is the middle-ground or middle-place, the "tiers passant" is the passing middle, the "tierce philosophie" the middle philosophy, etc. But the reader must keep in mind the image of the third which the excluded middle evokes in French, because it plays a decisive role in Serresean metaphorical thinking and may at times give it a slightly different slant than "middle" could effectively do.

In an interview with François-Bernard Huyghe, Serres talks about the tiers-instruit, translated as the "educated third party" (*The Unesco Courier* (December 1993): 4–7). For the purpose of preserving the fluctuating passage or fate of "exclusion" in Serres's work, from the parasite to the tiers-instruit, I have opted to use "instructed middle," as a derivation (or declination) of the excluded middle. I hope thereby to preserve the intimate relationship the French reader perceives between the terms "tiers exclu" and "tiers-instruit."

3. This piece enjoyed great popularity on the English stage in the late seventeenth century under the title "The Emperor of the Moon," thanks to free-wheeling changes the author Aphra Behn added to her translation of written scenarios into English. See Montague Summers, ed., *The Works of Aphra Behn* Vol. III (New York: Phaeton Press, 1967).

4. The relationship of postmodernism to science is discussed in detail by Thomas Weissert in an essay on Borges. He argues that chaos theory is a "move toward a postmodern physics in practice" (1991, 224). This observation is followed by a note in which he elaborates further: "I regard chaos theory as postmodern because . . . it abandons the requirement of a deterministic global theory of nonlinear dynamics in favor of a catalog of local theories. . . . But it is also true that the results of chaos theory, like those of quantum mechanics, are pressing the notion of scientific objectivity out of existence" (1991, 241n).

5. A curious and ambiguous commentary on harlequin as a national model (similar to the American Uncle Sam) is made in an old American encyclopedia published sometime after 1912, *The Americana: A Universal Reference Library* (New York: Scientific American Compiling Dept., ND). A short description of Harlequin that completely ignores its Italian roots, ends in the following statement: "The gallant, obsequious French harlequin is an entirely national mask."

6. This and all following quotes from *Le Tiers-Instruit* are my translation. The original French text reads: ". . . Tout est partout comme ici." (11)

7. Locally, of course, Harlequin is Serres confronting "press conferences" (critics and talk-show interviewers) where doubting scientists and other skeptics raise objections and ask questions in order to pinpoint his logic. Harlequin is also the French educational system synthesized in the figure of the teacher in front of his class, on whom high school students vent their frustration before taking to the streets, as happened frequently throughout 1990 and early 1991 (date of publication of *Le Tiers-Instruit*).

8. This scene and its interpretation by Serres are profoundly steeped in French tradition and mentality. How profoundly may be underscored by another playful setting, a sort of trivia game called "Roy qui ne ment" (The King Who Does Not Lie), popular among the nobles and courtiers of the thirteenth and fourteenth century. Cf. Ernst Hoepffner, "Frage- und Antwortspiele in der französischen Literatur des 14. Jahrhunderts," in *Zeitschrift für romanische Philologie* 33 (1909): 695–710. This game mirrors the sovereign's ethical authority as the center of truth and totalizing knowledge that distinguishes the ideal of French monarchy by divine right. Harlequin's boast of returning from his journey with the truth is directly related to this collective historical conscious.

9. "Même la peau d'Arlequin dément l'unité prétendue par son dire . . ." (p. 14)

10. The Cantor set and Koch curve are among the better known models scientists use to explain and visually represent the recursive process that lies at the heart of fractal geometry. Explanations and illustrations for the Cantor set can be found in Gleick, *Chaos*, p. 93; Hayles, *Chaos Bound*, pp. 156–57; and Benoît Mandelbrot, *The Fractal Geometry of Nature*, pp. 74–96. The latter uses the more precise term, Cantor's "middle third" which is especially close to Serres's neologism of the middle-instructed. Useful

explanations of the Koch curve can be found in Hayles, *Chaos Bound*, 166–67, and Gleick, *Chaos*, p. 99.

11. "the barren island of memory."

12. ". . . tous les sens se valent." Serres uses "le sens" for "direction," which is a deliberate choice that plays on the lexical fact that in French it also means "meaning" and "sense."

13. "Non, le jeu de pédagogie ne se joue point à deux, voyageur et destination, mais à trois. La tierce place intervient, là, en tant que seuil du passage. Or cette porte, ni l'élève ni l'initiateur n'en savent le plus souvent la place ni l'usage.

Un jour, à quelque moment, chacun passe par le milieu de ce fleuve blanc, état étrange du changement de phase, qu'on peut nommer sensibilité, mot qui signifie la possibilité ou la capacité en tous sens." (p. 29)

14. "Toute évolution et tout apprentissage exigent le passage par la tierce place. De sorte que la connaissance, pensée ou invention, ne cesse de passer de tierce en tierce place, s'expose donc toujours, ou que celui qui connaît, pense ou invente devient vite un tiers passant. Ni posé ni opposé, sans cesse exposé. . . . Voici déjà presque décrit le tiers-instruit dont l'instruction ne cesse pas: de nature et par ses expériences, il vient d'entrer dans le temps." (p. 34)

15. "Arlequin devient Pierrot." (p. 52)

16. A traditional linear reading would find an unresolved ambiguity in Serres's analogy, reinforced by the fact that two cognates are here intermingled (*kéntron* and *kéntrôn*), and that the assignation to the Greek word of a patchwork garment or poem pieced together from other works derives from rather obscure sources. I am grateful to my colleague Michael Johnson for pointing this out to me.

17. This playful linguistic creation occurs in the following context: "La recherche ou l'encyclopédie des connaissances, autrefois réputée ronde, . . . devient elliptique ou à deux foyers attractifs chez August Comte, déjà, par les sciences exactes et les sciences sociales, physique et sociologie, avant de se disperser aujourd'hui et de jouir, elle aussi, de plusieurs centres ou attracteurs; changent la forme et le concept de l'ancienne encyclopédie; on ne peut pas tout de même, la nommer *chaopédie*!" (p. 75)

18. ". . . la prescription admet l'action essentielle du temps." (213)

19. The German language reveals the same irrevocable link to time in its expression for prescription, "Verjährung" (Jahr = year): what is unlawful becomes "locally," or for a specific case, a law *under the effect of time*.

20. "En position tierce . . ." (p. 214)

21. "Le temps passe et ne s'écoule pas de façon passive; au contraire, il oublie ou efface les actes." (p. 213)

22. "On n'oublie pas l'oubli, acte en quelque sorte inoubliable." (p. 216)

23. Detailed discussions on mathematical operators can be found throughout *From Being to Becoming*. For a short, more accessible explanation, see Prigogine and Stengers, *Order out of Chaos*, p. 221–22.

24. "Pour les droits les plus positifs, nos actes se plongent dans le temps, mais pour la prescription, ils se font et se forment de temps, leur vraie matière première." (p. 215)

25. If the reader wishes to do so, a number may be plugged in for *x* corresponding to the number of occurrences of trope-stories up to this point in *Le Tiers-Instruit*. This

number would most probably have a different value for each reader inclined to do the computation, but the hypothetical discrepancy is symptomatic and perfectly natural for a dynamical or fractal reading.

26. "Je suis donc en réalité tous ceux que je suis dans et par les relations successives ou juxtaposées dans lesquelles je me trouve embarqué, . . . Donc le moi est un corps mêlé. . . . Voilà que revient le manteau d'Arlequin, cousu d'adjectifs, je veux dire de termes placés côte à côte." (p. 221)

27. ". . . la fluidité visqueuse" (p. 222)

28. ". . .mais de surcroît je suis sûr, absolument certain au-delà de toute espérance, qu'il existe un trou, une paille bizarre en ce panthéisme massif et dense, une exception étrange, source de toute douleur, que moi et *moi seul*, dans ce concert divin traversé de noise, ne suis pas Dieu; seulement cette faille de néant n'est pas Dieu; nouveau sens, très aigu, du vieux mot athée. Ici, pas de Dieu. Ici, seulement, Dieu s'absente. Ma part de destin est ce lieu d'athéisme.

Tout est Dieu sauf celui qui l'écrit, qui en lâche la plume parmi les pleurs." (pp. 230–31)

Serres borrows the image of an odd bit of straw from a sonnet by Paul Verlaine, discussed earlier in *Le Tiers-Instruit* (97), and uses it here in a sense that translates roughly into English as a "needle in a haystack."

29. ". . . mon métier consiste à écrire et à dire non point ce que je sais, ennuyeux, mort et passé, plus que parfait, mais, au contraire, ce que je ne sais pas et qui m'étonnera . . . l'imprévisible de l'artiste, l'inattendu et, rigoureusement, l'improbable." (p. 162)

6
Time and Earth: Reading *Le Contrat naturel*

1. This and subsequent quotes from Luc Ferry's study are my translation from the French original.

2. The translation is by Roger D. and Judith R. Masters, New York: St. Martin's Press, 1964. The original reads: "Le premier qui ayant enclos un terrain s'avisa de dire, *Ceci est à moi*, et trouva des gens assez simples pour le croire, fut le vrai fondateur de la société civile" (*Oeuvres complètes* III, 164).

3. The English biologist James Lovelock is most closely associated with the Gaia hypothesis. In his book, *The Ages of Gaia: A Biography of Our Living Earth* (New York, W. W. Norton and Co., 1988), he proposes an understanding of the earth as a discrete, single organism. Among American scientists, biologist Lynn Margulis is his staunchest supporter.

4. "Que de crimes, de guerres, de meurtres, que de misères et d'horreurs *n'eût* point épargnés au genre humain celui qui, arrachant les pieux ou comblant le fossé, *eût* crié à ses semblables: 'Gardez-vous d'écouter cet imposteur; vous êtes perdus si vous oubliez que les fruits sont à tous et que la terre n'est à personne!'" [emphasis added] (Discours sur l'origine . . . , *Oeuvres complètes* 164).

Epilogue

1. This short story from Balzac's collection *Scènes de la vie parisienne* became a well-known object of structuralist debate when R. Barthes analyzed it in *S/Z* (Paris: Seuil, 1970). In his text, Serres makes only fleeting reference to Barthes' study.

Bibliography

Michel Serres. *Hermès I. La communication.* Paris: Editions de Minuit, 1969.

———. *Hermès II. L'interférence.* Paris: Editions de Minuit, 1972.

———. *Hermès III. La traduction.* Paris: Editions de Minuit, 1974.

———. *Feux et signaux de brume. Zola.* Paris: Grasset, 1975.

———. *Hermès IV. La distribution.* Paris: Editions de Minuit, 1977.

———. *La naissance de la physique dans le texte de Lucrèce.* Paris: Editions de Minuit, 1977.

———. *Hermès V. Le passage du Nord-Ouest.* Paris: Editions de Minuit, 1980.

———. *The Parasite.* Trans. Lawrence R. Schehr. Baltimore: Johns Hopkins University Press, 1982. Trans. of *Le Parasite.* Paris: Grasset, 1980.

———. *Genesis.* Trans. Geneviève James and James Nielson. Ann Arbor: The University of Michigan Press, 1995. Trans. of *Genèse.* Paris: Grasset et Fasquelle, 1982.

———. *Rome: The Book of Foundations.* Trans. Felicia McCarren. Stanford, Calif.: Stanford University Press, 1991. Trans. of *Rome. Le livre des fondations.* Paris: Grasset 1983.

———. *Detachment.* Trans. Geneviève James and Raymond Federman. Athens, Ohio: Ohio University Press, 1989. Trans. of *Détachement.* Paris: Flammarion, 1983.

———. *Les Cinq Sens.* Paris: Grasset, 1985.

———. *L'Hermaphrodite.* Paris: Flammarion, 1987.

———. *Statues. Second livre des fondations.* Paris: François Bourin, 1987.

———. *The Natural Contract.* Trans. Elizabeth MacArthur and William Paulson. Ann Arbor: The University of Michigan Press, 1995. Trans. of *Le Contrat naturel.* Paris: François Bourin, 1990.

———. *Le Tiers-Instruit.* Paris: François Bourin, 1991.

———. *Angels: A Modern Myth.* Trans. Francis Cowper. Paris: Flammarion, 1993. Trans. of *La Légende des anges.* Paris: Flammarion, 1993.

———. *Les Origines de la géometrie. Tiers livre des fondations.* Paris: Flammarion, 1993.

Michel Serres with Bruno Latour. *Conversations on Science, Culture, and Time.* Trans. Roxanne Lapidus. Ann Arbor: The University of Michigan Press, 1995. Trans. of *Eclaircissements.* Paris: François Bourin, 1992.

Secondary Sources

Abraham, Ralph, and Terence McKenna. *Trialogues at the Edge of the West.* Sante Fe, New Mexico: Bear and Co., 1992.

Abraham, Ralph H., and Christopher D. Shaw. *Dynamics: The Geometry of Behavior.* Redwood City, Calif.: Addison-Wesley Publishing Co., 1992.

Argyros, Alexander J. *A Blessed Rage for Order.* Ann Arbor: The University of Michigan Press, 1991.

Balzac, Honoré de. *La Comédie Humaine. Vol 10. Etudes philosophiques.* Paris: Gallimard, 1979.

Bateson, Gregory. *Steps to an Ecology of Mind.* New York: Ballantine Books, 1972.

Baudelaire, Charles. *Les Fleurs du Mal et autres poèmes.* Paris: Garnier-Flammarion, 1964, p. 143.

Brady, Patrick. "Chaos Theory, Control Theory, And Literary Theory or: A Story of Three Butterflies." *Modern Language Studies* (Fall 1990): 65–79.

Brands, H. W. "Fractal History, or Clio and the Chaotics." *Diplomatic History* 16 (Fall 1992): 495–510.

Braverman, Richard. "Libertines and Parasites." *Restoration: Studies in English Literary Culture, 1660–1700* (Fall II, 1987): 73–86.

Coveney, Peter, and Roger Highfield. *The Arrow of Time: A Voyage through Science to solve time's greatest mystery.* London: W. H. Allen, 1990.

———. *Frontiers of Complexity: The Search for Order in a Chaotic World.* New York: Random House, 1995.

Crutchfield, James P., and J. Doyne Farmer, Norman H. Packard, Robert S. Shaw. "Chaos." *Scientific American* 255.6(December 1986): 46–57.

Culler, Jonathan. *On Deconstruction.* Ithaca, NY: Cornell University Press, 1982.

Curtius, Ernst Robert. *The Civilization of France.* Trans. Olive Wyon. New York: McMillan, 1932. Trans. of *Die Französische Kultur.* 2nd Ed. Bern: Francke Verlag, 1930.

Derrida, Jacques. *Dissemination.* Trans. Barbara Johnson. Chicago: The University of Chicago Press, 1981. Trans. of *La Dissémination.* Paris: Seuil, 1972.

Dubois, Monique, and Pierre Atten. "L'ordre chaotique." *La Recherche* 18.185 (February 1987): 190–201.

Ekeland, Ivar. *Mathematics and the Unexpected.* Chicago: The University of Chicago Press, 1988. Trans. by Ivar Ekeland of *Le Calcul, L'Imprévu. Les figures du temps de Kepler à Thom.* Paris: Seuil, 1984.

Feigenbaum, Mitchell J. "Universal Behavior in Nonlinear Systems." *Los Alamos Science* (Summer 1980): 4–27.

Ferenczi, Thomas. "*Les cinq sens,* de Michel Serres." Review. *Le Monde* 15 (November 1985): 27–30.

Ferry, Luc. *Le nouvel ordre écologique.* Paris: Grasset, 1992.

———, reviewer. "Maîtres penseurs et apprentis." *L'Express* 20 (April 1990): 60–61.

Girard, René. *Deceit, Desire, and the Novel.* Baltimore: Johns Hopkins University Press, 1965.

———. *Violence and the Sacred.* Baltimore: Johns Hopkins University Press, 1977.

———. *Things Hidden since the Foundation of the World*. Trans. Stephen Bann and Michael Meteer. London: Athlone Press, 1987. Trans. of *Des choses cachées depuis la fondation du monde*. Paris: Grasset, 1978.

Gleick, James. *Chaos: Making a New Science*. New York: Viking Penguin, 1987.

Gray, Richard T. "The Dialectic of 'Enscentment': Patrick Süskind's *Das Parfum* as Critical History of Enlightenment Culture." PMLA 108.3 (May 1993): 489–505.

Greening, Marcella. "Strange Attractors and Novelistic Discourse." SLS Conference, Montreal, 1991.

Greimas, A.-J. *De l'imperfection*. Périgneux, France: Pierre Fanlac, 1987.

Giusti, Ada. "L'Académie française et l'évolution de la langue. Entretien avec Michel Serres." *Contemporary French Civilization* 20, No. 1 (Winter/Spring 1996), 106–122.

Harari, Josué V., and David F. Bell. *Hermes: Literature, Science, Philosophy*. Baltimore: Johns Hopkins University Press, 1982.

Hawking, Stephen W. *A Brief History of Time*. New York: Bantam Books, 1988.

Hayles, N. Katherine. *Chaos Bound: Orderly Disorder in Contemporary Literature and Science*. Ithaca, NY: Cornell University Press, 1990.

Hertz-Ohmes, Peter. "Serres and Deleuze: Hermes and Humor." *Canadian Review of Comparative Literature* 14, No. 2 (June 1987): 239–250.

Huizinga, J. *The Waning of the Middle Ages*. Garden City, NY: Doubleday and Co., 1954.

James, Geneviève. "Entretien avec Michel Serres." *The French Review* 60.6 (May 1987): 788–796.

The Jerusalem Bible. Garden City, NY: Doubleday and Co., 1966.

Kauffman, Stuart. *At Home in the Universe: The Search for the Laws of Self-Organization and Complexity*. New York: Oxford University Press, 1995.

Kavanagh, Thomas M. "Michel Serres, *Les cinq sens*." Book Discussion. *MLN* 101.4 (September 1986): 937–941.

Kellert, Stephen H. *In the Wake of Chaos: Unpredictable Order in Dynamical Systems*. Chicago: The University of Chicago Press, 1993.

Krauthammer, Charles. "Saving Nature, But Only for Man." *Time* 17 (June 1991): 82.

Latour, Bruno. "The Enlightenment Without the Critique: A Word on Michel Serres' Philosophy." *Contemporary French Philosophy*. Ed. A. Phillips Griffiths. New York: Cambridge University Press, 1987, pp. 83–97.

———. "Pasteur on Lactic Acid Yeast: A Partial Semiotic Analysis." *Configurations* 1 (1992): 129–145.

Leitch, Vincent B. *Deconstructive Criticism*. New York: Columbia University Press, 1983.

Lewin, Roger. *Complexity: Life at the Edge of Chaos*. New York: McMillan Publishing Co., 1994.

Lloyd, Genevieve. *The Man of Reason: "Male" and "Female" in Western Philosophy*. Minneapolis: University of Minnesota Press, 1984.

Mandelbrot, Benoît. *The Fractal Geometry of Nature*. New York: W. H. Freeman and Co., 1983.

May, Robert. "Simple mathematical models with very complicated dynamics." *Nature* 261 (June 10, 1976): 459–67.

Paulson. William R. "Literature, Knowledge, and Cultural Ecology." *SubStance* 22, No. 2 and 3 (1993): 27–37. Madison: The University of Wisconsin Press.

———. *The Noise of Culture*. Ithaca, NY: Cornell University Press, 1988.

———. "Le Cousin Parasite: Balzac, Serres et le démon de Maxwell." *Stanford French Review* 9, No. 3 (Winter 1985): 397–414.

Peitgen, H.-O., and P. H. Richter. *The Beauty of Fractals: Images of Complex Dynamical Systems*. New York: Springer-Verlag, 1986.

Prigogine, Ilya, and Isabelle Stengers. *Order out of Chaos*. Boulder, Colorado: Shambhala Publications, Inc., 1984.

———. *La nouvelle alliance*. Paris: Gallimard, 1979, 1986 (for preface to 2nd. ed.).

Prigogine, Ilya. *From Being to Becoming*. San Francisco: W. H. Freeman and Co., 1980.

Reeves, Hubert. *Malicorne: Réflexions d'un observateur de la nature*. Paris: Seuil, 1990.

Ricoeur, Paul. *Time and Narrative*. Trans. Kathleen Blamey and David Pellauer. Vol III. Chicago: The University of Chicago Press, 1988. Trans. of *Temps et récit III*. Paris: Seuil, 1985.

Rilke, Rainer Maria. *The Notebooks of Malte Laurids Brigge*. Trans. Stephen Mitchell. New York: Random House, 1982. Trans. of *Die Aufzeichnungen des Malte Laurids Brigge. Saemtliche Werke VI*. Frankfurt a/M: Insel-Verlag, 1966.

Rousseau, Jean-Jacques. *The First and Second Discourses*. Trans. Roger D. and Judith R. Masters. New York: St. Martin's Press, 1964. Trans. of *Discours sur l'origine, et les fondements de l'inégalité parmi les hommes. Oeuvres complètes III*. Paris: Gallimard, 1964.

Stengers, Isabelle. "Les Régimes de la vérité." *Psychanalystes. Revue du Collège de Psychanalystes* 44 (Fall 1992): 9–28.

Storey, Robert F. *Pierrot: A Critical History of a Mask*. Princeton: Princeton University Press, 1978.

Süskind, Patrick. *Perfume: The Story of a Murderer*. Trans. John E. Woods. New York: Pocket, 1987. Trans. of *Das Parfüm: Die Geschichte eines Mörders*. Zurich: Diogenes, 1985.

Verne, Jules. *From the Earth to the Moon. Around the Moon*. Trans. Edward Roth. New York: Dover Publications, 1962.

Weissert, Thomas P. "Dynamical Discourse Theory." *Time and Society* 4, No. 1 (February 1995): 111–133.

———. "Representation and Bifurcation: Borges's Garden of Chaos Dynamics." In *Chaos and Order*. Edited by N. Katherine Hayles. Chicago: The University of Chicago Press, 1991, pp. 223–43.

———. "Conjugating Motion," in *Textures of Time*. Edited by Paul Harris. Ann Arbor: The University of Michigan Press, publication forthcoming.

Index